CANCER BIOLOGY AND BIOSYNTHESIS

CANCER BIOLOGY AND BIOSYNTHESIS

Edited by

Samuel H. Wilson

A volume in honor of
Mahlon B. Hoagland

CRC Press
Boca Raton Ann Arbor Boston

Library of Congress Cataloging-in-Publication Data

Cancer biology and biosynthesis / edited by Samuel H. Wilson.
p. cm.
Includes bibliographical references.
Includes index.
ISBN 0-936923-50-4
1. Cancer--Congresses. 2. Cancer cells--Congresses. 3. Cell transformation--Congresses. 4. Biosynthesis--Congresses. 5. Hoagland, Mahlon B.--Congresses
I. Wilson, Samuel H., 1939- .
[DNLM: 1. Cell Transformations, Neoplastic. 2. Melanins--physiology. 3. Melanoma. 4. Neoplasms--analysis. 5. Transcription, Genetic. QZ 200 C2151103]
RC261.A2C363 1991
616.99'4'071--dc20
DNLM/DLC
For Library of Congress 90-11291
CIP

Developer: Telford Press

Direct all inquiries to CRC Press, Inc., 2000 Corporate Blvd., N.W., Boca Raton, Florida, 33431.

International Standard Book Number 0-936923-50-4

Library of Congress Card Number 90-11291

Printed in the United States

Contents

Preface

We now see that one of the fundamental tenets of of molecular biology is that amino acid residues themselves cannot interpret the genetic code embodied in messenger RNA. Instead, each amino acid relies upon an adapter molecule, termed transfer RNA, for this most critical decoding step in the flow of the genetic iinformation froma gene to a protein molecule. Mahlon Hoagland was instrumental in the experiments behind these statements, having discovered the aminoacyl-tRNA charging reaction in 1953 and transfer RNA, itself, within the subsequent three years.

The conceptual simplicity and coherence of the decoding process by transfer RNA is truly one of nature's wonders and one that Mahlon Hoagland and Paul Zamecnik were privileged to enjoy before the rest of us. This volume is in honor of Mahlon, who worked alongside several of the contributors (M.B., J.L., P.Z) and was a mentor to the others. In preparing this Preface, I considered it appropriate to briefly chronicle the amino acid activation and transfer RNA story. Yet, this job has been done recently by none other than Mahlon Hoagland himself, in a commemorative article for BBA. This article is reproduced here in its entirety. The first paper on tRNA was published by Mahlon and associates in BBA as a "note," and the detailed follow-up article was published in JBC in 1957.

Samuel Wilson
Bethesda, MD

Dr. Mahlon B. Hoagland
Photograph courtesy of WFEB archives. Sculpture by Lawrence Beale Smith.

Mahlon B. Hoagland

I first became aware of the challenge of the problem of protein biosynthesis in 1949 at the Huntington Laboratories at the Massachusetts General Hospital, Boston. Paul Zamecnik and a small group of associates were then pioneering the use of ^{14}C-labeled amino acids to explore protein synthesis in mammalian cells. They were progressing from in vivo systems to tissue slices and thence to the earliest cell-free preparations. They were at pains to show that incorporation of isotopes into protein represented true peptide bond formation, and they had established that the process was coupled to oxidative energy production.

I had just finished medical school after a 2-year bout with tuberculosis and had got a postdoctoral position at the Huntington with the thought of returning to medicine in a year or two. I worked on the carcinogenic and other biological effects of the metal beryllium - a project that convinced me I wanted to continue to do science - and over the next 2 years became increasingly interested in Zamecnik's work. He encouraged me to get some in-depth exposure to protein chemistry and to biochemical energetics. I spent a year with Kaj Linderstrom-Lang at the Carlsberg Laboratory in Copenhagen and another year with Fritz Lipmann at the Biochemical Research Laboratories back at the MGH. My exposure to the latter's interest in the ATP-dependent activation of acetate was an important factor in leading me to my discovery of amino-acid activation in 1954, soon after my return to the Huntington. By then the Zamecnik group had delineated the protein synthesizing system of rat liver as consisting of microsomes (ribosomes), a soluble fraction (in which I had found the activating enzymes), ATP and ATP generating system, and GTP

In ruminating upon the activation reaction, Zamecnik wondered if the enzymes involved might also participate in the synthesis of RNA. Could adenyl-amino acids be, in a sense, double-barreled - donating their amino acids to the protein polymerizing machinery or, alternatively, their adenyl portion to a nucleotide polymerizing system? To test the idea, Zamecnik and Mary L Stephenson, his long-term associate at the Huntington, incubated ^{14}C-labeled ATP with ribosomes and soluble fraction and looked to see if RNA became labeled. It did. (This phenomenon was later shown by Zamecnik and Lisalotte Hecht to be due to a special reaction: the terminal adenylation of transfer RNA.) They were pleased with the result, but suspicious: could the AMP moiety of ATP, or indeed the whole ATP molecule, be attached to RNA in some trivial, non-specific, non-covalent linkage? They did a control experiment incubating the same system with ^{14}C-labeled amino acids and found to their surprise that they, too, became bound to RNA. Furthermore, when the two cell fractions were separately incubated with labeled amino acids the soluble fraction labeled its indigenous RNA startlingly more actively than the ribosomal fraction labeled its RNA.

At this point - January 1956 - Zamecnik stopped doing experiments. He'd just been appointed as Joseph Aub's successor as head of the Huntington Laboratories and Professor of Oncologic Medicine at Harvard. He also had clinical duties in the hospital which put additional demands on his time. In June, Zamecnik told me of the 5-month-old findings. I was enormously intrigued by what was clearly a direct interaction between amino acids and RNA - between the building units of protein and what we presumed to be information-bearing molecules in the very fraction that I had found carried out the first step of protein synthesis. I pursued the lead.

The existence of a small amount of low-molecular-weight RNA in cells had already been noted by that time. It was assumed to be a product of the breakdown of ribosomal RNA during the preparation of cell fractions. Here, however, was unexpected evidence that 'soluble RNA' might have some very special function. It seemed likely that the same enzymes that were responsible for activating amino acids by acylating the adenyl portion of ATP were transferring the amino acyl groups to soluble RNA. This was borne out by my early studies reported in the cited BBA paper: the aminoacylation of soluble RNA was specifically dependent on ATP (GTP, CTP and UTP wouldn't do); the addition of pyrophosphate caused a rapid deacylation of the RNA (presumably by pushing the reaction to the left to form ATP and free amino acid); and the product was acid-stable and alkali-labile, did not contain free amino acid, and produced a hyroxamic acid when incubated with hydroxylamine.

The exciting broader interpretation of the phenomenon was that we had found in this small amino-acid-binding RNA a key intermediate in protein synthesis - a molecule or molecules that accepted activated amino acids and carried them to the ribosomes. (Other possibilities - that the binding was an *in vitro* artifact or that the RNA was some kind of a waste disposal unit picking up unwanted amino acids - were sufficiently unattractive to be set aside.) That such a hypothetical intermediate might have a role in the mysterious process of ordering amino acids, we had no doubt - and no evidence. The crucial first test was obvious: see if amino acids could be transferred from soluble RNA to peptide linkage in protein in the absence of ATP.

In the most suspenseful and exciting few hours of my professional life, I did an experiment that came out essentially as reported in the cited paper: having labeled the indigenous RNA of the soluble fraction with amino acids in the presence of ATP, reisolated the fraction's protein and RNA free of amino acids and ATP by reprecipitation at pH 5, I then incubated it with microsomes. To my delight, most of the counts on RNA were rapidly transferred to protein, in a reaction that specifically required GTP and proceeded in the presence of a large excess of free unlabelled amino acid.

Stephenson, Zamecnik and I spent the remaining months of 1956 and early 1957 pinning down the evidence that 'soluble RNA' was indeed a key intermediate in protein synthesis. ('Soluble' RNA was soon to be named transfer RNA (tRNA) as its function became better established.) We confirmed the transfer's dependence on GTP and showed the transfer to be rapid (complete in 5 min at 37 degrees C) compared to the overall rate of incorporation of free amino acids into protein. We separated the transfer RNA with attached amino acids by phenol extraction from the soluble fraction, added fresh soluble fraction and GTP and showed that the amino acids were quantitatively transferred to protein on ribosomes. And we showed that the same events occurred using ribosomes from cell-free preparations of mouse ascites tumor cells.

Importantly, too, we observed that as we added increasing quantities of different amino acids to the system there was room for all of them. This had to mean that each of the 20 amino acids was separately bound to some site on the RNA molecules. The RNA must then, either be a single species with binding sites for all 20 of the amino acids or 20 (or more) different RNAs, each with its own specific binding site for each amino acid. The latter, of course, proved to be the case but we couldn't make the distinction then. We were convinced we had uncovered the second major step in the cell's machinery for making protein.

I wrote up and submitted the cited BBA account in late 1956. By that time, T. Hultin in Sweden, was obtaining indirect evidence that amino acids passed into a state experimentally distinguishable from free amino acid before they entered protein. Kikuo Ogata in Japan, and Paul Berg and Robert Holley in the U.S.A. were also finding signs of steps in which amino acids were bound to RNA after activation. A more definitive account of our work appeared in 1958 (1).

What justified our inference that transfer RNA had anything to do with determining order? The evidence was then only circumstantial: if DNA carried genetic information specifying amino-acid sequence, and if RNA carried out DNA's role in the cytoplasm - which everyone by then assumed it must - it followed that the observed direct chemical association of amino acids with RNA must have implications for sequencing.

In late 1956, I had my first visit from a card-carrying 'molecular biologist'. (We at the Huntington considered ourselves biochemists.) Jim Watson had just become Professor of Biology at Harvard and was probing the structure of the ribosome. He had heard rumors of our discovery of transfer RNA and I jubilantly told him of our findings. He was restlessly attentive and when I'd finished he told me that Francis Crick had forecast the existence of transfer RNA-like molecules! Hadn't I heard of the adaptor hypothesis? I was astonished and admitted that I hadn't. Watson explained that Crick had been wrestling with the question: if the machinery of protein synthesis contained information specifying the order of amino acids in protein - presumably in the form of DNA-derived RNA acting as the ordering template - how would the template "know" - i.e., chemically recognize - an amino acid if it bumped into it? This concern was a natural extension into chemistry - into structure - of Crick's fascination with the genetic code. There wasn't a trace of chemical homology or complementarity between amino acids on the one hand, and RNA on the other. Suppose, thought Crick, an enzyme first attached each of the amino acids to a special, unique short length of RNA. The amino acid would then have had conferred upon it a chemical identity that could be recognized by an ordering RNS template by hydrogen bonding between complimentarily shaped bases. Crick envisioned that the postulated small molecules, which he called adaptors - in that they adapted the amino acids so as to be recognizable by an ordering template - would be really quite small, perhaps only 3 to 10 bases in length. According to coding theory, three would be the minimum number of bases needed to code for a single amino acid. (The theory glossed over the problem of how the system would 'know' how to attach the amino acid to the correct RNA adaptor - vaguely relegating the job to the versatility of an enzyme).

I was bowled over by the ingenuity and beauty of the idea and sensed it had to be the explanation of our experimental findings. An image arose in my mind; we biochemical explorers slashing our way through a dense jungle to discover a beautiful temple while Francis Crick, floating gracefully overhead on gossamer wings of theory, waited patiently for us to see the goal that he was already gazing down upon!

REFERENCE

1. Hoagland, M.B., Stephenson, M.L., Scott, J.F., Hecht, L.I. and Zamecnik, C. (1958) *J. Biol. Chem.* **231**: 241-257.

Contributors

Nancy L.R. Bucher
Department of Pathology
Boston University School of Medicine
80 East Concord Street
Boston, Massachusetts 02118

Marc H. Dresden
Verna and Marrs McLean
Department of Biochemistry
Baylor College of Medicine
Houston, Texas 77030

Robert Engel
Doctoral Programs in Chemistry and Biochemistry
The City University of New York
Queens College
Flushing, New York 11367

Barry Gotlinsky
PAL Industries
73 Bloomingdale Road
Hicksville, NY 11801

Helene Z. Hill
Section of Cancer Biology
Department of Radiology
New Jersey Medical School
Newark, New Jersey 07103

John W. Littlefield
Department of Physiology
Johns Hopkins University School of Medicine
Baltimore, Maryland 21205

Thomas S. Parker
Department of Biochemistry
Cornell Medical Center
429 E. 71st St.
New York, NY, 10021

R. V. Quincey
The University of Birmingham
School of Biochemistry
The University of Birmingham
P.O. Box 363
Birmingham, England B15 2TT

Oscar A. Scornik
Department of Biochemistry
Dartmouth Medical School
Hanover, NH 03756

Burton E. Tropp
Doctoral Programs in Chemistry and Biochemistry
The City University of New York
Queens College
Flushing, New York 11367

Samuel H. Wilson
Laboratory of Biochemistry
National Cancer Institute
National Institutes of Health
Bethesda, MD 20892

Paul Zamecnik
Worchester Foundation for Experimental Biology
222 Maple Avenue
Shrewsbury, MA 01545

Foreword

One day in 1948, fresh from medical school, Mahlon Hoagland turned up at the Huntington Laboratories and offered to work on cancer research if there was a place for him. A number of us were so engaged, the term cancer research being loosely defined under the gentle eye of Joseph Aub, its Director. Mahlon began to study the mechanism of beryllium toxicity. The addition of beryllium to the phosphors of fluorescent lights had induced in plant workers exposed to it an unusual syndrome of weakness, breathlessness, and lymphadenopathy related to inhalation of beryllium-containing dusts. Several publications emerged from this investigation, pinpointing the site of action of beryllium. During the course of these studies, Mahlon became increasingly fascinated by the field of protein synthesis, just beginning to send up shoots and to unfold little flowers in the laboratory which surrounded him. He determined to join this group, but to get a better grip on biochemistry, he thought a period of immersion in a good laboratory elsewhere would be essential. Thus he spent a year in the Carlsberg Laboratories in Copenhagen in 1951-1952. At that moment, however, Lindstrom-Lang, its gifted Director, had lost interest in science, and was often painting, playing the violin or just smoking long cigars, while the cause of science languished temporarily. There was much more excitement in the Kalckar laboratory across town, which was in those years a mecca, attracting among others young scientists such as Paul Berg, Jim Watson and Hans Klenow.

One year was not, however, enough for this transformation from medical training poised for surgery to the molecular knife of biochemistry. Mahlon returned to the laboratory of Fritz Lipmann, our distinguished neighbor at the Massachusetts General Hospital, where he remained from 1952-1954. In a foreshadowing experiment with David Novelli he found as part of an ATP-requiring step in coenzyme A synthesis that AMP and pyrophosphate were generated.

As Mahlon had hoped, he then returned to the Huntington Laboratories, and joined the quickening chase for an understanding of the amino acid activation step in protein synthesis. A cell-free protein synthesizing system had just been found, which was crying out for dissection of the precise role of the ATP ribosomal fraction of the liver cell homogenate or with the soluble proteins. Our bets were on the former, as a way of ordering the sequence of protein at the moment of amino acid activation, whatever the mechanism might be. The results clearly showed, however, that the ribosomes were not required for amino acid activation, but that the soluble protein fraction was. Emerging from this discovery was a fact hard to swallow at the time, that a separate activating enzyme must exist for each amino acid, in the reaction ATP + aa + E AMP-aa.E + pp. This was a landmark discovery, which opened the door to modern protein synthesis. There were soon hints that a component was missing from the picture of activated amino acid interacting with a ribosome, the latter composed of protein and ribonucleic acid. Transfer RNA, or as we first designated it, soluble RNA, proved to be the missing ingredient.

In a classic experiment, Mahlon showed in a system composed of ^{14}C-aminoacyl-AMP, soluble activation enzymes and ribosomes, that as the aminoacyl group left aminoacyl-AMP it was built into a nascent protein chain associated with the ribosomes. Thus, the essential role of transfer RNA in protein synthesis was elucidated and became firmly established. It did not escape Mahlon's attention that the translation step in protein synthesis was thus the interaction of an anticodon segment of transfer RNA with a codon piece of RNA associated with the ribosome. Experimental clothing was in this way nicely supplied to the contemporaneous theoretical idea of Francis Crick that a trinucleotide adaptor must exist to translate the message of the gene into that of the protein.

Three short years encompassed the discovery of two enormous building stones of modern biochemistry plus the development of the critical adaptor hypothesis. Certainly enough for the epitaph of a scientist, but there were other important features in Mahlon's career - - for one, his conversion of the Worcester Foundation for Experimental Biology from a center for reproductive physiology into an exciting gathering place for molecular biologists. As the founder and initial spokesman for the Delegation for Basic Science, he also pleaded the cause of research successfully in the halls of Congress. He has as well written lucidly about science, with the tongue of a central participant but in a language comprehensive to the intelligent generalist.

For these varied and intertwined reasons, in this publication from his friends we salute Mahlon's accomplishments and his influence on those of us whose lives he has for the better directly touched.

Paul Zamecnik

1

GROWTH SIGNALS ARE WHERE IT STARTS

Nancy L. R. Bucher

Department of Pathology
Boston University School of Medicine
Boston, MA 02118

INTRODUCTION

Mahlon Hoagland, whom this volume honors, in his early days at the Huntington Laboratories, used to entertain us with his guitar, and a ballad about ... "the ankle bone's connected to the shin bone, the shin bone's connected to the knee bone ..." and so on, until the whole thing got put together. This chapter, which deals with the signals that control growth of normal adult tissues is connected in similar fashion to the subject of this book. Growth is connected to DNA connected to RNA connected to protein synthesis; the problem is how overall regulation of these processes is implemented by signals from the body.

Growth regulation has been investigated extensively in systems ranging from unicellular organisms, where cell division is largely a function of nutrient availability, to higher forms where cells proliferate in concert with the needs of the organism as a whole. In mammals, complex homeostatic mechanisms impose tight controls, maintaining fluctuations of pH, temperature, and the chemical environment within narrow physiological limits. Within these constraints, extracellular growth signals, which may be positive, negative, or an interplay of both, appear to modulate genetically determined growth of specific tissues in accordance with functional demands. These signals comprise blood borne factors acting singly or in synergistic combinations, facilitated by nutrients whose intracellular concentrations may themselves be hormonally regulated. Specificity of the signals for a given type of cell may be conferred by the particular factor or combination of factors appropriate for the metabolic state and available receptor sites on the plasma membrane. Activation of the coordinated programs of gene replication that are associated with cell division may occur through mechanisms that are different for different cell types, or even for the same type under different conditions. This is the big picture, needing to be fleshed out at the molecular level.

RAT LIVER AS A MODEL

A familiar model, in the quest for the signals that regulate growth of normal, differentiated tissues is rat liver. Within the adult animal, hepatocytes, though active metabolically, are quiescent in terms of growth, and may be regarded as *in vivo* prototypes of cells in the so called G_0 phase of the cell cycle - - *i.e.*, in a non-cycling state, generally requiring extracellular signals for reactivation of growth, and manifesting a prolonged transit time through the initial prereplicative period (G_1 phase). DNA synthesis marks the end of G_1 and entry into S-phase. (Some investigators, however, view G_0 not as a separate phase, but merely an extreme prolongation of G_1.) In general, extracellular growth controls act on processes in the G_0/G_1 phase of the cell cycle. The expression of DNA synthesis, therefore, indicates that the cell has re-engaged in the programmed cell cycle process, which yields two daughter cells, generally without further intervention of external stimuli (1,2,3). In the rat, excision of the two main liver lobes, comprising slightly over 2/3 of the whole organ, induces an abrupt outburst of DNA synthesis, starting after a 16 hour prereplicative phase and reaching a peak at 22-24 hours. Mitosis follows a similar course 6 to 8 hours later. Although the excised lobes do not regrow, restoration of the original liver mass through enlargement of the remnant occurs mostly within a week, with gradual reversion to quiescence. Liver growth is also induced, though not as dramatically, in response to less drastic tissue deficits, resulting from physical, chemical, or toxic injury, or to relative functional deficiencies imposed on the intact organ by metabolic or hormonal imbalances. These may be physiological, as in pregnancy (4) or aberrant , as in diabetes (5). Excision of less than the usual 2/3 of the liver results in rises in DNA synthesis proportional to the size of the liver loss; small deficits are repaired slowly, despite the liver's large growth capacity (6). This adjustment of rate of repair in proportion to liver deficit is also evident in the continually diminishing proliferative activity seen during the course of regeneration following partial hepatectomy.

Autoradiographs following pulse labeling with ^{3}H thymidine show that the initial peak of DNA synthesis in regenerating liver represents exclusively hepatocyte DNA replication, since bile duct, littoral, endothelial and other cell types do not become labeled until a day or two later. If the ^{3}H-thymidine is administered continuously during the first 2-3 days, nearly all of the hepatocyte nuclei become labeled; therefore virtually all hepatocytes are potentially proliferative. Because the hepatocytes are arrested in G_0 they must replicate their DNA before undergoing mitosis; failure of cytokinesis in some of the cells results in an actual increase in polyploidy in the regenerating liver. An obvious but frequently overlooked point is that within the animal growth signals are specific for a given cell type, yet most of the known mitogens are effective in a variety of cells. (For reviews see 2,7,8,9,10,11,12,13,)

GROWTH SIGNALS: WHOLE ANIMAL STUDIES

Years of study in whole animals have failed to answer the basic questions: what mechanisms initiate the growth process, adjust its rate to comply with the liver deficit, and terminate it when the original tissue complement is restored? Nevertheless a number of informative observations have emerged, some of which are paralleled in more recent studies in primary hepatocyte cultures. There is now general agreement that control is exerted by blood borne signals, based on early studies showing a growth response to partial hepatectomy in liver transplants (14), or more definitively in normal livers of rats following cross circulation of blood with a partially hepatectomized partner. This blood exchange resulted in transmission of a growth signal to the normal liver of the intact partner that was proportional to the size of the liver deficit created by the partial hepatectomy (15). Moreover, it was necessary for the cross circulation to continue for at least 12-14 hours for the cells to become irreversibly committed to synthesize DNA, although a 2-4 hour hiatus did not diminish the commitment, suggesting a stepwise progression (16). Parallel behavior is observed in cell cultures where relatively long exposure to a mitogen is required for commitment of a significant number of cells to synthesize DNA, even though binding of growth factors to plasma membrane receptors is rapid (17). Also, a stepwise progression is consonant with the concept of "commitment" and progression through the cell cycle, demonstrable in some cultures and discussed later.

The metabolic state of the cells strongly influences their response to a regenerative stimulus. DNA synthesis following partial hepatectomy was significantly modulated by the feeding schedule. Moreover, merely by manipulation of protein or amino acid intake, a large, though transient elevation of DNA synthesis was induced (18,19), whereas a similar protocol in pre-fasted rats yielded a negative result (20). To what extent these effects are due to the nutrient down-and up- shifts, or to the associated hormonal imbalances, cannot be deduced from *in vivo* studies.

Evidence seeming to implicate a number of hormones as growth regulators has not been definitive. Pancreatic hormones in particular are highly responsive to nutrient availability, and insulin and glucagon seemed among the more promising hormones to explore. Divergent reports from other laboratories led us eventually to study splanchnic eviscerated rats (*i.e.* animals surgically deprived of the entire gastrointestinal tract, pancreas and spleen). Partial hepatectomy in these eviscerated rats caused a rise in DNA synthesis, which although delayed and greatly diminished compared to partially hepatectomized normal rats, was highly significant when compared to eviscerated, non-hepatectomized control animals (21). The low rate of DNA synthesis in the partially hepatectomized eviscerated animals was restored to normal by infusion of insulin and glucagon in combination, but not by either hormone alone (22). Likewise, these two hormones together but not separately, strikingly increased survival of mice infected with large doses of A59 mouse virus which caused a 100% fatal fulminant hepatitis in untreated control animals (23). These experiments, involving 2 different animal models, brought to light the importance of synergistic interactions among hormones and growth regulating factors.

That the same insulin and glucagon treatment when applied to normal animals failed to initiate DNA synthesis, however, suggested that these hormones had a permissive role and that in addition a mitogenic agent was required (24). This notion was further supported by the finding that in eviscerated animals, lacking pancreatic hormones, low but significant regenerative activity was demonstrable, indicating the presence of mitogenic stimulation by one or more substances not of pancreatic or splanchnic origin (21).

Epidermal growth factor, the mouse equivalent of the human hormone urogastrone, is a potent mitogen for a broad spectrum of cells. Continuous intraperitoneal infusion of this substance into normal rats induced a small but significant increase in DNA synthesis and mitosis which was potently amplified by large but physiological doses of insulin. Glucagon also augmented the EGF stimulus, but only at excessive dose levels (25,26). Glucagon and insulin together tended to inhibit the EGF stimulus, whereas the two hormones had a mutually reinforcing, positive action in the evisceration and A-59 virus infection experiments already described, in which the rats were in a severely stressed condition. We previously observed that prestressed rats responded differently than normal rats to partial hepatectomy (27). It appears that synergistic combinations of extracellular factors required for growth regulation may vary depending upon the prior state of the cells.

The transient stimulation of hepatocyte proliferation by the EGF-insulin combination occurred in the presence of the full complement of liver cells, overriding the normal endogenous regulatory mechanisms. Similarly, in animals with severe streptozotocin-induced diabetes, whose livers were already modestly enlarged, insulin treatment precipitated an early surge in DNA synthesis. As the diabetes became controlled by the continuing insulin treatment, however, the total hepatic DNA content, an index of the excess hepatic cell complement, gradually returned to a normal level. This observation points to the existence of a mechanism that acts to restore normal cellularity when an overproduction of cells has occurred (5).

In the experiments just described (25), the increased DNA synthesis resulting from the intraperitoneal infusion of EGF and insulin occurred only in the liver. That this indicated a specificity for hepatocytes is uncertain, as EGF is a potent mitogen for many types of cells; probably the amounts of infused hormone reaching other tissues were negligible, as the major portion would be absorbed into the portal circulation and be carried directly to the liver which is known to take up nearly all of the EGF and much of the insulin as well (28,29).

Other hormones that, like insulin and glucagon, influence hepatic carbohydrate metabolism, include vasopressin and the adrenergic hormones, which also modulate liver growth. In the Brattleboro rat, a mutant strain congenitally lacking vasopressin, hepatic regeneration was found to be impaired, but was restored to near normal by infusion of the missing hormone (30). Adrenergic hormones are suggested to have a role in liver growth by several investigators (31,32,33). Vasopressin and alpha$_1$ adrenergic agonists are also effective in hepatocyte cultures, as described later.

On the basis of many studies, the liver has been shown to regenerate, generally to a lesser extent than normal, in the complete absence of the pituitary, thyroid, parathyroid, or adrenal cortex; conversely, the hormones secreted by these glands have been reported to influence growth of normal and regenerating livers, in varying degrees, often shown only with pharmacological doses. (10). The specific hormones tested include pituitary growth hormone, somatomedin C, triiodothyronine, parathyroid hormone, calcitonin, prostaglandins and several stimulatory or inhibitory substances purified or partially characterized from serum. The variability of the results reported suggests that interactions with other hormones or groups of hormones may be a factor. The multiplicity of agents seems bewildering in view of instances in other tissues of direct hormone/target-organ feedback mechanisms. That signals for liver growth should exhibit a higher order of complexity may not be surprising, however, in view of the liver's vast array of activities. Multiple, variable signals may be necessary to adjust growth to so wide a range of functional needs.

The experimental work described so far was carried out exclusively at the level of the whole animal, before suitable methods had been developed to permit study of hepatocytes in culture. Despite the limitations of the *in vivo* approach, it has enabled the regenerative process to be characterized, and allowed delineation of aspects of growth regulation whose mechanisms must be deciphered at the molecular level. The synergy among growth factors, seen as important within the animal, has become an established concept, much emphasized in current cell culture models for growth control, mainly in fibroblasts (3,34).

HEPATOCYTE CULTURES

During recent years methods have been developed for isolation of adult rat hepatocytes, virtually free of other cell types, by perfusion of the liver with collagenase (35,36). The hepatocytes are maintained as monolayer cultures in media composed only of purified chemicals and hormones, without serum — rigorous conditions permitting dissection of mechanisms at the molecular level in ways impossible within the animal.

Soon after the freshly isolated hepatocytes are plated they flatten, spread, and form confluent aggregates. There is a transient induction of "heat shock" proteins (37). In 2-3 days, even in the absence of growth factors, DNA synthesis rises modestly and occasional mitotic figures are seen. These activities although low are significantly above those occurring *in vivo*. Therefore, due merely to isolation and culture, an unknown number of hepatocytes move out of the G_0 state, and at least some progress completely through the cell cycle in the absence of serum or external growth factor stimulation. This mobilization could be due to nutrient down- and up-shifts, or cell surface perturbations during the isolation process, or, more likely, loss of cell-cell and cell-matrix contacts. After a few days signs of "dedifferentiation" are seen, manifested by reversion towards a fetal phenotype (38,39,40). On the other hand, dedifferentiation appears to be minimal or absent during hepatic regeneration *in vivo* (41,42,43).

A number of factors proposed as regulators of liver growth within the animal stimulate hepatocyte proliferation in these cultures. In early studies hepatocytes survived but did not actively proliferate. In 1976 Friedman and his associates (44,45) showed that DNA synthesis could be stimulated by EGF, especially if insulin, glucagon, or both were added, but mitotic figures were initially seldom seen. Very high rates of DNA synthesis and abundant mitoses were obtained soon thereafter by various modifications of culture conditions, such as inclusion of high concentrations of metabolic substrates (especially pyruvate) (46,47), reduced calcium concentration, (46) and lowered cell population density (48).

Like glucagon, vasopressin (49) and adrenergic hormones (50,44) were found also to augment the stimulus afforded by EGF and insulin, under suitable experimental conditions. The augmenting effect of these three hormones may be due, at least in part, to stimulation of glycogenolysis, a function shared by all three hormones, which may provide intracellular substrates necessary for full expression of the growth promoting effects of other agents (51,52). This action would be consonant with the considerable amplification of DNA synthesis seen when growth factors are supplemented with high concentrations of pyruvate (53). These observations are in accord with the whole animal studies, and similarly point to the importance of the metabolic state of the cells in relation to growth stimuli.

An implication of the synergy among the growth factors seen both *in vivo* and *in vitro* is that under physiological conditions within the animal, small shifts in balance among the various interacting growth signals could greatly amplify a subliminal proliferative stimulus.

SIGNAL TRANSDUCTION

According to current concepts (51,54), binding of vasopressin, glucagon or alpha-1-adrenergic agonists to their respective hepatic type plasma membrane receptors initiates a rapid sequence of reactions involving turnover of phosphatidylinositols (PI) with release of diacylglycerol (DAG) and a series of inositol phosphates (IP's), some of which cause a rise in intracellular Ca^{2+}. DAG is an important source of free arachidonate which gives rise to the eicosanoids, including prostaglandins, thromboxanes and others. The production of "second messengers" is the important feature of this system whose intricate ramifications are still under investigation. The second messengers are DAG, inositol phosphates, Ca^{2+} and various eicosanoids. All appear to function as activators of protein kinases, especially protein kinase C, although they may function in other ways as well (51,55). The protein kinases, in turn, regulate cellular activities through phosphorylation of various enzymatic, structural and regulatory proteins. It remains unclear how the external signals from various hormones, that are implemented through activation of the PI system, can cause a diverse array of end functions. Mitchell (55) concludes "the point that emerges particularly forcefully from all of these recent findings is that the metabolism of inositol phosphates in stimulated cells is startlingly complex" — which seems an understatement; the brief description presented in this paragraph is highly oversimplified. Although the PI system has been well studied in liver, how it operates in regulation of liver growth remains unclear.

Receptor binding of beta-adrenergic agonists, does not affect the PI system, but instead activates adenylate cyclase, causing a rise in cyclic AMP. On the contrary, binding of $alpha_2$-agonists depresses cyclic AMP levels, through mechanisms that are less well established (56). An extensive literature documents regulation of various metabolic activities of hepatocytes by cyclic AMP dependent protein kinases. The adrenergic hormones epinephrine and norepinephrine bind to both alpha- and beta-type receptors, although norepinephrine is predominantly an alpha agonist and epinephrine predominantly a beta agonist. In regenerating liver *in vivo* and in hepatocytes cultured *in vitro* there is a relatively early shift from alpha- to beta-type adrenergic receptors (57,58,59,60,61). The stimulation of DNA synthesis by norepinephrine in the hepatocyte cultures was shown by Michalopoulos and his associates to be due to its interaction with the alpha$_1$-type receptor; no significant increases were found with alpha$_2$- or beta-adrenergic agonists, at least when the agents were added at the time the cells were plated (50).

In the case of glucagon, hepatocytes have two types of receptors, the GR-1 type functioning through the PI second messenger system, and the GR-2 type through activation of adenylate cyclase. High glucagon concentrations tend to promote cyclic AMP formation and inhibit the PI pathway (62).

Insulin presents another variant, exerting its effect through cleavage of a novel inositol-glycolipid with release of an IP-glycan and a myristic acid-containing DAG that differs from the well known arachidonic acid-containing DAG discussed previously. The IP-glycan functions as a second messenger to regulate cyclic AMP phosphodiesterase (63).

It is clear from the foregoing that both cyclic AMP and PI turnover are involved in signaling a number of cellular functions; only some of these, however, lead to cell proliferation. Reviews by Whitfield and Michell and their associates (51,64,65,) discuss roles for these substances in growth.

In a recent collaborative study (66) we found that following vasopressin binding the receptor/ligand complex was rapidly internalized, and within a few minutes the receptor recycled to the cell surface. Vasopressin binding to hepatocytes cultured 3-5 hours after cell isolation, and binding to cells in livers perfused *in situ* were similar with respect to receptor dynamics. In the hepatocyte cultures, during the first 48 hours as DNA synthesis became active, there was a progressive loss of receptors from the cell surface with a concomitant elevation in intracellular receptors, although the overall number of receptors decreased. Similar phenomena have been reported in regenerating liver with respect to glucagon, EGF and asialoglycoprotein receptors (67,68,69); a reduction in surface number and increase in internal pools of these receptors were found at 17-24 hours (but not at 7 hours) after partial hepatectomy. These increases in internal receptor pools roughly coincided with the rise in DNA synthesis *in vivo*, which as noted earlier occurs at 15-24 hours (66). Despite a number of striking parallels in behavior between hepatocytes cultured *in vitro* and liver regenerating *in vivo*, there are major differences.

THE EXTRACELLULAR MATRIX, CELL-CELL INTERACTIONS, THE CYTOSKELETON AND GENE EXPRESSION

Darnell and associates in studies of gene expression in normal and regenerating mouse liver *in vivo* and in hepatocytes in short-term cultures (41,70,71), have found that genes relating to functions specific to liver are actively expressed, control in normal liver operating mainly at the level of transcription. On the other hand, expression of cytoskeletal and other genes common to all cells was low and relatively constant, and control was post-transcriptional, due largely to prolongation of mRNA half-life. During liver regeneration most liver-specific mRNA's showed little change in concentration or transcription rate, whereas cytoskeletal mRNA's increased up to 10-fold without an equivalent transcriptional increase. Many of the constituents necessary to increase cell mass may be supplied by increased preservation and utilization of transcribed mRNA's. It has previously been shown that also at the translational level, the accumulation of protein in the growing liver remnant is largely attributable to a near cessation of protein breakdown rather than to substantial increases in synthesis (72,73). It appears that the differentiated adult hepatocyte continues to function as a differentiated cell during the 1 or 2 cell divisions necessary for complete hepatic regeneration (41,42,43).

Darnell and associates also found that in hepatocyte cultures within 24 hours there was an 80-99% drop in transcription of most liver-specific mRNA's (11 or more were tested), but not of cytoskeletal mRNA's. A wide variety of culture conditions failed to prevent this loss, but slices of liver in which liver structure was temporarily destroyed and reconstituted did exhibit transcription of liver-specific genes (70).

Recent efforts in several laboratories to preserve normal hepatocyte attributes in cultures have met with some degree of success. Guguen-Guillouzo and associates have studied co-cultures of hepatocytes and a type of non-parenchymal liver epithelial cells (RLEC's), which are thought to be derived from terminal bile ductules. The hepatocytes became enmeshed in reticulin fibers, survived for prolonged periods, and both transcribed and expressed a number of liver-specific genes. There was, however, no DNA synthesis. It appears that for maintenance of normal liver functions, in addition to soluble hormones, growth factors and nutrients, a critical role should be assigned to cell-cell interactions, both between hepatocytes themselves as well as between hepatocytes and RLEC's, which promote development of an adequate ECM (38,74,75).

Reid and associates (76,77) have focused upon the effects of both hormones and the extracellular matrix in hepatocyte cultures, with recent emphasis on purified matrix components. Although none of numerous hormones and growth factors tested could restore transcription of liver-specific genes in hepatocyte cultures, mRNA half-lives were substantially prolonged when the medium was supplemented with 5 hormones. On purified Type IV collagen in the presence of this supplement, hepatocytes exhibited sustained growth over a 4-week period, compared to little or none on purified Type I collagen. Transcription, however, remained low, even when laminin and fibronectin were added. On the other hand, addition of certain glycosaminoglycans (GAGs) or proteoglycans (PGs) to the supplemented cultures variously stimulated expression of liver-specific genes at both transcriptional and post-transcriptional levels, while at the same time reducing expression of cytoskeletal genes entirely by post-transcriptional mechanisms. Thus, uniquely among all substances previously investigated, these PG's and GAG's, which are ECM components, restored the normal *in vivo* program of high liver-specific and low cytoskeletal gene expression, although activation of liver specific genes was variable. With respect to cell-cell communication, these investigators found that gap junctions, which disappeared after 24-hours in culture even in the presence of the hormone supplement, were restored to varying degrees by the GAG's, PG's, and anionic polysaccharides — especially by heparin and dermatan sulfate proteoglycan (76).

The studies of Bissell and associates are also concerned with the ECM. In the most recent report (78), hepatocytes were cultured in the absence of serum or hormones on a gel matrix extracted from the Engelbreth-Holm-Swarm mouse tumor (EHS), described by Kleinman, Martin and associates (79). This tumor produces basal lamina/extracellular matrix in exceptional abundance. The morphology of these cultures differed strikingly from the usual flattened monolayer; cells remained compact and rounded, with minimal spreading, forming mounded aggregates in thick rope like strands. This appearance persisted, along with normal hepatocyte ultrastructure, for at least 3 weeks in medium without growth factors or serum. Albumin secretion remained stable at *in vivo* levels, and several other liver functions persisted for a similar period. By immunofluorescent techniques it was shown that within the normal liver there is a basal lamina on the sinusoidal surface of the hepatocyte that contains Type IV collagen, heparan sulfate proteoglycan, laminin and also fibronectin, which may not be an integral component. None of these substances, when isolated from the EHS tumor and added separately, could effectively substitute for the complete EHS gel matrix; full support of hepatocellular functions appears to require a complex of matrix proteins (78).

Our recent collaborative studies with S.R. Farmer and associates are consonant with and extend the foregoing reports. They show that hepatocytes cultured on EHS matrix exhibit a program of gene expression resembling the normal adult liver. This is quite opposite from the program in hepatocytes cultured on gels prepared from rat tail collagen (mainly Type I), whose cytoskeletal protein and mRNA content are abnormally high while mRNA's of liver specific genes are attenuated. On EHS gels, moreover, DNA synthesis is substantially depressed both in normal cells and in pre-stimulated hepatocytes isolated from regenerating livers, although a limited degree of response to growth factor stimulation is maintained. These results apply to cells plated at low or medium cell densities, whereas DNA synthesis on rat tail collagen is very active in the presence of growth factors. At high density a different picture emerges; DNA synthesis is strongly inhibited on both rat tail and EHS matrices (80,81). Clarification of how physiological modulation of liver growth relates to functional demands will require culture conditions promoting maintenance of hepatocytes in G_0 in the presence of normal cell-cell contacts, with full expression of liver-specific genes and responsiveness to hormonal/growth factor stimulation. More definitive knowledge of cell-matrix and cell-cell interactions seems essential.

Mounting evidence, based upon several cell types, suggests that hormonal signals require for their implementation an appropriate cell morphology, which in turn depends upon the cell's interaction with its extracellular matrix and with neighboring cells, as well as the organization of its internal cytoskeleton. Experimental alteration of cell shape induces changes in gene expression, manifested, for example, by changes in mRNA's and synthesis of specific proteins (82,83,84,85,86). Within the animal, in many tissues including liver, changes in cell shape of comparable magnitude are not evident, and if present may be minimal due to constraints of tissue structure, while cytoskeletal remodeling may still be taking place. We are currently investigating this problem both *in vivo* and in hepatocyte cultures. Our preliminary studies in cultures confirm our early observation that in partially hepatectomized rats tubulin synthesis and DNA synthesis, which increase rapidly at about the same time after growth stimulation, are not tightly coupled; inhibition of DNA synthesis with injections of araC left tubulin synthesis unaltered (87,88).

CONCLUSIONS

The picture that emerges is one of hierarchical controls with multiple levels of complexity. External signals including growth factors act in synergistic combinations through binding to receptors on the cell surface which are subject to modulating influences. Signal transduction from the cell surface occurs *via* elaborate interacting second messenger systems. Cell responses to external signals are modulated by their contacts with other cells and with the ECM, which may exert its influence in various ways including alterations in the cell surface or cytoskeletal organization. Whatever the mechanisms, the end result is regulation of gene expression by growth signals, but within the constraints imposed by cell-cell and cell-matrix interactions.

The discovery of oncogenes, and the normal genes from which they are derived, has revealed that these are the genes whose proteins are the growth factors operating in normally regulated cell growth. What is needed is further understanding of how the signal generated by the binding of the growth factor to its receptor at the cell surface leads ultimately to cell division (89). Although the generality of this problem applies to all cells, the liver may superimpose its own special rules. Without a clearer grasp of the fundamental mechanisms through which external influences operate — whether they be soluble substances such as hormones, growth factors and nutrients, or insoluble factors such as external matrices and structural components of cell-cell contacts — we will be unable to point with certainty to the physiological mechanisms through which hepatocyte proliferation is controlled, and why the signals are specific for the liver.

ACKNOWLEDGEMENT

The work from my laboratory reviewed here has extended over many years and received support from USPHS Grants CA02146, CA39099 and AM194351, the American Cancer Society, the Damon Runyon Memorial Fund for Cancer Research, Inc., and very recently the Council for Tobacco Research.

REFERENCES

1. Baserga, R. (1985). *The Biology of Cell Reproduction*, Harvard University Press, Cambridge.
2. Ord, M. G., and Stocken, L. A. (1984). *Cell and Tissue Regeneration*, John Wiley & Sons (Eds.) New York.
3. Pledger, W. J., Meymandi, A., Mitchell, P., and Olashaw, N. (1987). **In**: *Molecular Mechanisms in the Regulation of Cell Behavior* (Waymouth, C., Ed.) A. R. Liss, Inc. New York pp. 19-24.
4. Campbell, R. M., Fell, B. F., and Mackie, W. S. (1974). *J. Physiol.*, **241**:699-713.
5. Steiner, D. F. (1978). **In**: *Hepatotrophic Factors*. Ciba Foundation Symposium no. 55 (New Series) Elsevier, New York pp. 229-236.
6. Bucher, N. L. R., and Swaffield, M. N (1964). *Cancer Res.*, **24**:1611-1625.

7. Alison, M. R. (1986). *Physiol. Rev.*, **66**:499-541.
8. Bucher, N. L. R. (1987). **In**: *The Isolated Hepatocyte: Use in Toxicology and Xenobiotic Transformations* (Rauckman, E. J., and Padilla, G. M., Eds.) pp. 1-19 Academic Press, New York.
9. Bucher, N. L. R., and Malt, R. A. (1971). *Regeneration of Liver and Kidney*, Monograph in New Eng. J. Med, Medical Progress Series pp. 1-278, Little, Brown & Co., Boston, MA.
10. Bucher, N. L. R., and McGowan, J. A. (1985). **In**: *Liver and Biliary Disease: A Pathophysiological Approach* (Wright, R., Alberti, G., Karran, S., and Millward-Salder, H. Eds.) pp. 210-227, W.B. Saunders Co., Ltd., London.
11. Fausto, N. (1984). *Mol. Cell. Biochem.*, **59**:131-147.
12. Leffert, H. L., Koch, K. S., Lad, P. J., Skelly, H., and deHemptinne, B. (1982*a*). **In**: *Hepatology* (Zakim, D., and Boyer, T. D. Eds.) pp. 64-75, Saunders, Philadelphia.
13. Leffert, H. L., Koch, K. S., Lad, P. J., Skelly, H., and deHemptinne, B. (1982*b*). **In**: *The Liver: Biology and Pathobiology* (Arias, I. M., Popper, H., Schachter, D., and Schafritz, D. A., Eds.) pp. 601-614, Raven, N.Y.
14. Leong, G. F., Grisham, J. W., Hole, B. V., and Albright, M. L. (1964). *Cancer Res.* **24**:1496-1501.
15. Moolten, F. L., and Bucher, N. L. R. (1967). *Science*, **158**:272-274.
16. Bucher, N. L. R., Schrock, T. R., and Moolten, F. L. (1969). *Johns Hopkins Med. J.*, **125**:150-257.
17. Carpenter, G., Stotscheck, C. M., and Soderquist, A. M. (1982). *Ann. New York Acad. Sci.*, **397**:11-17.
18. Leduc, E. H. (1949). *Am. J. Anat.*, **84**:397-421.
19. Short, J., Armstrong, N. B., Kolisky, M. A., Mitchell, R. A., Zemel, R., and Lieberman, I. (1974). **In**: *Control of Proliferation in Animal Cells* (Clarkson, B., and Baserga, R., Eds.) pp. 37-48, Cold Spring Harbor Laboratory, Cold Spring Harbor, N.Y.
20. Bucher, N. L. R., McGowan, J. A., and Patel, U. (1978*a*). **In**: *Cell Reproduction vol. XII* (Dirkson, E. R., Prescott, D. M., and Fox, C. F., Eds.) pp. 661-670, ICN/UCLA Symposia on Molecular and Cellular Biology, Academic Press, New York.
21. Bucher, N. L. R., and Swaffield, M. N. (1973). *Can. Res.*, **33**:3189-3194.
22. Bucher, N. L. R., and Swaffield, M. N. (1975*a*). *Proc. Natl. Acad. Sci. USA*, **72**:1157-1160.
23. Farivar, M., Wands, J. D., Isselbacher, K. J., and Bucher, N. L. R. (1976). *New Engl. J. Med.*, **295**:1517-1519.
24. Bucher, N. L. R., and Swaffield, M. N. (1975*b*). **In**: *Advances in Enzyme Regulation* vol. 13 (Weber, G., Ed.) pp. 281-293, Pergamon Press, Inc., New York.
25. Bucher, N. L. R., Patel, U., and Cohen, S. (1978*b*). **In**: *Hepatotrophic Factors*, Ciba Foundation Symposium no. 55 (new series) pp. 95-107, Elsevier/Excerpta Medica/North Holland, New York.
26. Bucher, N. L. R., Patel, U., and Cohen, S. (1978*c*). **In**: *Advances in Enzyme Regulation* (Weber, G., Ed.), **16**:205-213, Pergamon Press, Inc., New York & Oxford.
27. Moolten, F. L., Oakman, N. J., and Bucher, N. L. R. (1970). *Canc. Res.*, **30**:2353-2357.
28. Caruana, J. A., Goldman, J. K., Camara, D. S., and Gage, A. A. (1981). *Surg. Gynecol. Obstet.*, **153**:726-730.
29. St. Hilaire, R. J., Hradek, G. T., and Jones, A. L. (1983). *Proc. Natl. Acad. Sci. USA*, **80**:3797-3801.
30. Russell, W. E., and Bucher, N. L. R. (1983). *Am. J. Physiol.*, **245**:G321-G324.
31. Hasegawa, K., and Koga, M. (1977). *Life Sci.*, **21**:1723-1728.
32. Morley, C. (1981). *Int. J. Biochem.*, **13**:969-973.
33. Cruise, J. L., Knechtle, S. J., Bollinger, R. R., Kuhn, C., and Michalopoulos, G. (1987). *Hepatol.*, **7**:11849-11894.
34. Rozengurt, E. (1986). *Science*, **234**:161-166.
35. Berry, M. N., and Friend, D. S. (1969). *J. Cell Biol.*, **43**:506-520.

36. Seglen, P. O. (1976). **In**: *Methods in Cell Biology* (Prescott, D. M., Ed.), **13**:29-83, Academic Press, New York.
37. Wolffe, A. P., and Tata, J. R. (1984). *FEBS Lett.*, **176**:8-15.
38. Guguen-Guillouzo, C. (1986). **In**: *Isolated and Cultured Hepatocytes* (Guillouzo, A., and Guguen-Guillouzo, Eds.) pp. 240-283, Eurotext/INSERM, Paris.
39. Leffert, H., Moran, T., Sell, S., Kelley, H. , Ibsen, K., Mueller, M., and Arias, I. (1978). *Proc. Natl. Acad. Sci USA*, **75**:1834-1838.
40. Sirica, A. E., Richards, W., Tsukada, Y., Sattler, C. A., and Pitot, H. C. (1974). *Proc. Natl. Acad. Sci. USA*, **76**:283-287.
41. Friedman, J. M., Chung, E. Y., and Darnell, J. E., Jr. (1984). *J. Mol. Biol.*, **179**:37-53.
42. Huber, B. E., Heilman, C. A., Wirth, P. J., Miller, M. H., and Thorgeirsson, S. S. (1986). *Hepatology*, **6**:209-219.
43. Petropoulos, C. J., Lemire, J. M., Goldman, D., and Fausto, N. (1985). *Cancer Res.*, **45**:5114-5121.
44. Friedman, D., Claus, T. H., Pilkis, S., and Pine, G. (1981). *Exp. Cell Res.* **135**:283-290.
45. Richman, R. A., Claus, T. H., Pilkis, S. S., and Friedman, D. L. (1976). *Proc. Natl. Acad. Sci. USA*, **73**:3589-3593.
46. Hasegawa, K., Watanabe, K., and Koga, M. (1982). *Biochem. Biophys. Res. Comm.*, **104**:259-265.
47. McGowan, J. A., and Bucher, N. L. R. (1983). **In**: *Isolation, Characterization and Use of Hepatocytes* (Harris, R. A., and Cornell, N. W., Eds.) pp. 165-170, Elsevier, New York.
48. Ichihara, A., Nakamura, T., and Tanaka, K. (1982). *Mol. Cell Biochem.*, **43**:145-160.
49. Russell, W. E., and Bucher, N. L. R. (1983*b*). **In**: *Isolation, Characterization and Use of Hepatocytes* (Harris, R. A., and Cornell, N. W., Eds.) pp. 171-176, Elsevier, New York.
50. Cruise, J. L., Houck, K. A., and Michalopoulos, G. K. (1985). *Science*, **227**:749-751.
51. Downes, C. P., and Michell, R. H. (1985). **In**: *Molecular Mechanisms of Transmembrane Signalling* (Cohen, P., and Houslay, M., Eds.), **4**:3-35, Elsevier, New York.
52. Verity, M. A., Travis, G., and Brown, W. J. (1972). *Lab. Invest.*, **27**:108-114.
53. McGowan, J. A., and Bucher, N. L. R. (1983). *In Vitro* **19**:159-166.
54. Greenberg, M. E., and Ziff, E. B. (1984). *Nature*, **311**:433-438.
55. Michell, R. H. (1986). *Nature*, **319**:176-177.
56. Cerione, R. A., Regan, J. W., Nakata, H., Codina, J., Benovic, J. L., Gierschik, P., Somers, R. L., Spiegel, A. M. , Birnbaumer, L., Lefkowtiz, R. G., and Caron, M. G. (1986). *J. Biol. Chem.*, **261**:3901-3909.
57. Aggerbeck, M., Ferry, N., Zafrani, E. S., Billou, M. C., Barouki, R., and Hanoune, J. (1983). *J. Clin. Invest.*, **71**:476-486.
58. Bronstad, G., and Christoffersen, T. (1980). *FEBS Lett.*, **120**:89-93.
59. Nakamura, T., Tomomura, A., Kato, S., Noda, C., and Ichihara, A. (1984). *J. Biochem.*, **96**:127-136.
60. Okajima, F., and Ui, M. (1982). *Arch. Biochem. Biophys.*, **213**:658-668.
61. Refsnes, M., Sandnes, D., Melien, O., Sand, T. E., Jacobsen, S., and Christoffersen, T. (1983). *FEBS Lett.*, **164**:291-298.
62. Wakelam, M. J. O, Murphy, G. J., Hruby, V. J., and Houslay, M. D. (1986). *Nature*, **323**:68-71.
63. Berridge, M. J. (1987). *Ann. Rev. Biochem.*, **56**:159-193.
64. Boynton, A. L., and Whitfield, J. F. (1983). *Adv. Cyclic Nucleotide Res.*, **15**:193-294.
65. Whitfield, J. F., Sikorska, M., Bossi, D., Boynton, A. L., Durkin, J. P., Kleine, L., Rixon, R. H., and Youdale, T. (1986). **In**: *Intracellular Calcium Regulation* (Bader, H., Gietzen, K., Rosenthal, J. Rudel, R., and Wolf, H. V., Eds.) pp 179-202, Manchester University Press, U.K.
66. Fishman, J. B., Dickey, B. F., Bucher, N. L. R., and Fine, R. E. (1985). *J. Biol. Chem.*, **260**:12641-12646.
67. Bhargava and Levine (1982). *Gastroenterology*, **82**:1005 -.

68. Howard, D. J., Stockert, R. J., and Morell, A. G. (1982). *J. Biol. Chem.*, **257**:2856-2858.
69. Leffert, H., Alexander, N. M., Faloona, G., Rubalcava, B., and Unger, R. (1975). *Proc. Natl. Acad. Sci. USA*, **72**:4033-4036.
70. Clayton, D. F., Harrelson, A. L., and Darnell, J. E., Jr. (1985). *Mol. Cell Biol.*, **5**:2623-2632.
71. Powell, D. J., Friedman, J. M., Ouellette, A. J., Krauter, K. S., and Darnell, J. E., Jr. (1984). *J. Mol. Biol.*, **179**:21-35.
72. Augustine, S. L., and Swick, R. W. (1980). *Am. J. Physiol.*, **238**:E46-52.
73. Scornik, O. A. (1972). *Biochem. Biophys. Res. Commun.*, **47**:1063-1066.
74. Fraslin, J. M., Kneip, B., Vaulong, S., Glaise, D., Munnich, A., and Guguen-Guillouzo, C. (1985). *EMBO J.*, **4**:2487-2491.
75. Guguen-Guillouzo, C., Clement, B., Baffet, G., Beaumont, C., Morel-Chang, E., Glaise, D., and Guillouzo, A. (1983). *Exper. Cell Res.*, **143**:47-54.
76. Fujita, M., Spray, D. C., Choi, H., Saez, J., Watanabe, T., Rosenberg, L. C., Hertaberg, E. L., and Reid, L. M. (1987). *Hepatology*, **7**:1S-9S.
77. Reid, L. M., Narita, M., Fujita, M., Murray, Z., Liverpool, C., and Rosenberg, L (1986). **In**: *Isolated and Cultured Hepatocytes* (Guillouzo, A., and Guguen-Guillouzo, C., Eds.) pp. 225-258, Eurotext/INSERM, Paris.
78. Bissell, D. M., Arenson, D. M., Maher, J. J., and Roll, F. J. (1987). *J. Clin. Invest.*, **79**:1-11.
79. Kleinman, H. K., McGarvey, M. L., Hassell, J. R., Star, V., Cannon, F. B., Laurie, G. W., and Martin, G. R. (1986). *Biochem.*, **25**:312-318.
80. Ben Ze'ev, A., Robinson, G. S., Bucher, N. L. R., Farmer, S. R. (1987). *J. Cell Biol.*, **105**:50a (abstract).
81. Ben Ze'ev, A., Robinson, G. S., Bucher, N. L. R. and Farmer, S. R. (1988) *Proc. Natl. Acad. Sci. USA* **85**: 2161-2165.
82. Ben Ze'ev, A., Farmer, S., and Penman, S. (1980). *Cell*, **21**:365-372.
83. Bissell, M. J., Hall, H. G., and Parry, G. (1982). *J. Theor. Biol.*, **99**:31-68.
84. Farmer, S., Ben Ze'ev, Benecke, B., Penman, S. (1978). *Cell*, **15**:627-637.
85. Folkman, J., and Moscona, A. (1978). *Nature*, **273**:345-349.
86. Ingber, D. E., Madri, J. A., and Folkman, J. (1987). **In:** *Vitro Cell and Develop. Biol.*, 23:387-394.
87. Bucher, N. L. R., and Berkley, P. (1975). **In**: *The Cell Cycle in Malignancy and Immunity*, 13th Hanford Biology Symposium (Hampton, J. C., Ed.) pp. 67-75, Richland, Washington Technical Information Center USERDA.
88. Robinson, G. S., Bucher, N. L. R., and Farmer, S. R. (1986). *J. Cell Biol.*, **103**:274a (abstract).
89. Newmark, P. (1985). *Nature*, **317**:380.

2

COMPACTION AND DIFFERENTIATION OF EMBRYONAL CARCINOMA CELL AGGREGATES

John W. Littlefield

Department of Physiology
Johns Hopkins University School of Medicine
Baltimore, MD 21205

The behavior of aggregates of embryonal carcinoma cells is of interest primarily because it seems to resemble closely events which occur at the morula stage in the mammalian embryo and shortly thereafter (1,2). The blastomeres of the mouse embryo appear morphologically identical up to the early 8-cell stage, at which point they form a grape-like cluster. Soon thereafter, the cells have maximized the apposition of their cell membranes, giving a smooth outer surface to the morula in which the outlines of individual cells are less distinct. This process is called "compaction" (3,4). It appears to be preceded briefly in time by the initiation of gap junctions between cells and "polarization" of the constituent blastomeres (5,6), although the exact temporal relationship between these processes is not yet certain (7). In polarization the organelles and cytoskeleton of the blastomeres become oriented asymmetrically, and the apical and basolateral surfaces differentiate (8-11). Both cell coupling and polarization seem to begin without extensive cell apposition, so they may be the first morphological signs of compaction (5,12).

Compaction of the 8-cell embryo allows for the creation of two populations of cells, external and internal. The apical surfaces of the former make up the external surface of the morula, while the latter cells are entirely surrounded by the external cells. Subsequently, the external cells form the trophectoderm of the blastocyst, while the internal cells become the inner cell mass (ICM). Trophectoderm cells are soon terminally differentiated, while ICM cells go on to form the embryo proper, plus other tissues. The formation of trophectoderm from the external cells of the compacted morula reflects the first "commitment" and differentiation of the embryo. The second leads to the formation in similar fashion of "primitive" endoderm cells from the surface of the ICM which is exposed to the blastocoele cavity.

Thus the formation of an external layer of cells which soon differentiates, in contrast to internal cells which remain undifferentiated, is an important mechanism in early embryogenesis, and perhaps again later on, for example, in somite formation. How does the external location of cells determine their biochemical fate? In the fertilized eggs of some lower organisms the distribution of organelles is asymmetrical, so that their allocation to daughter cells is uneven. Presumably this is what leads to the different fates of these daughter cells and their progeny. However, there is no evidence for this mechanism in mammalian eggs. It was first suggested that the external *location* of certain cells directed their differentiation, perhaps by exposure to a different microenvironment (8,13-14). In contemporary terms this could mean exposure to higher or lower concentrations of the paracrine factors which we now know to participate in the regulation of growth and differentiation. More recently, it has been suggested that the polarization of the cells of the morula creates a situation in which the next cell division, which almost always occurs in a plane parallel to the surface of the morula, separates the outward-looking apical end from the inner basal end of each cell (11,15). These ends contain different assortments of organelles and other constituents, so that the resulting daughter cells are different, which somehow leads to the differentiation of trophectoderm. In addition to these two well-known conceptions of how differentiation at the morula stage begins, others are possible, and two others are mentioned below.

Much is known already about the process of compaction and subsequent differentiation in the early mouse embryo (3,4,16). But the amount of material available is limited and the embryo is difficult to access for experimental manipulation. This has lead to the development of techniques for the culture of embryos *in vitro*, as well as the derivation of cell cultures from embryos. Such cultures can provide much more material and can be manipulated with the techniques of somatic cell genetics. The first cultures derived from embryos, and still the most commonly studied, originated from embryonic tumors (teratocarcinomas) arising spontaneously or following transfer of a mouse embryo to an ectopic site (17,18). These cultures are called embryonal carcinoma cell (ECC) lines. In recent years it has become possible to derive cultures directly from blastocysts (19-21). These cultures, which are called ES cell lines, hold considerable promise for future research (22), although they do form tumors when injected into mice, just like other primary mouse cell cultures (23).

Mouse ECC are believed to resemble embryonic ectodermal cells of the ICM, because they never produce trophectoderm but usually can differentiate into endoderm-like cells, either spontaneously or when exposed to retinoic acid or certain other agents. This process seems analogous to the second commitment and differentiation of the early mouse embryo mentioned above, that is, the formation of primitive endoderm from the ICM. The appearance of several biochemical markers characteristic of endoderm have been examined in a number of ECC lines, including those markers which distinguish parietal and visceral endoderm (2). Different ECC lines usually differentiate to one or the other type of endoderm, but a few can be made to change direction. Differentiation is usually terminal after a few cell divisions, but on occasion partially differentiated cells continue to grow and divide. Also when one or more cells of certain ECC lines are injected into a blastocyst, they participate subsequently in the formation of various differentiated tissues (24).

For some ECC and ES cell lines, the differentiation to endoderm-like cells also occurs spontaneously if the cells are allowed to aggregate in suspension. The first detailed studies of such differentiated aggregates (called "embryoid bodies") were carried out by Martin and colleagues (25-27). They showed that clonal sublines of the pluripotent teratocarcinoma line called PSA1, which is usually maintained on a monolayer of feeder cells, formed tight rounded aggregates when grown without this monolayer. These aggregates soon acquired a distinct outer layer of cells with the morphological characteristics of endoderm, including the formation of a basement membrane underneath the cells. In contrast to undifferentiated embryonal carcinoma cells, this endodermal layer contained very little alkaline phosphatase activity. Some of these aggregates, especially from certain sublines, went on to develop central fluid-filled cavities reminiscent of the blastocyst. If the embryoid bodies were allowed to attach again to a surface, extensive differentiation occurred in the outgrowth of cells over the ensuing weeks. Martin and colleagues were the first to suggest that this behavior overall resembled the formation of primitive endoderm from the ICM and to point out the advantages of this system as a model for embryogenesis (1,25).

Next, Hogan, Taylor and Adamson (28) showed that another well-known embryonal carcinoma cell line called F9, which shows little spontaneous differentiation, could be induced to form a peripheral layer of visceral endoderm if F9 aggregates were treated with a low concentration of retinoic acid. Previously it had been shown that monolayer cultures of this line formed parietal endoderm when exposed to a somewhat higher level of retinoic acid plus cyclic AMP (29). Grover, Oshima, and Adamson (30) analyzed the F9 aggregate system in considerable detail. Evidence was provided for six sequential stages of aggregation, proliferation, differentiation, organization, maturation, and secretion, based on immunofluorescent analyses of the appearance of several endoderm markers over several days, as well as immunoprecipitation of secreted proteins. Without retinoic acid the aggregates developed diffuse staining for extracellular matrix components, such as laminin and type IV collagen, but when exposed to 5×10^{-9} M retinoic acid these components were located mainly underneath a peripheral layer of endoderm. A few days later this peripheral layer produced alpha-fetoprotein, a characteristic of visceral endoderm. When exposed to a higher concentration of retinoic acid in combination with cyclic AMP, the differentiated external layer was no longer evident, alpha-fetoprotein was not produced, and the staining for extracellular components was again diffuse throughout the aggregate.

In a following paper Grover, Andrews, and Adamson (31) suggested a key role for laminin in the basement membrane in stimulating the formation of the overlying endoderm layer. Subsequently these investigators focused upon the roles of other extracellular matrix components, such as type IV collagen and fibronectin (32,33). Then they confirmed the earlier suggestion (29) that F9 stem cells first differentiate to primitive endoderm, which is next directed to either parietal or visceral endoderm depending on the concentrations of retinoic acid and cyclic AMP (34,35). Also, they found that interconversion from visceral to parietal endoderm cells could be produced up to a certain point in time. A recurrent theme throughout this work has been the directive role of the basement membrane, probably especially its laminin, in the formation of the external layer of visceral endoderm - a theme which has been stressed also by others (36,37).

We have been studying another ECC line called H6 (38), which was derived from a CBA mouse embryo placed in an ectopic site. When exposed in monolayer culture to retinoic acid, this line differentiates into endoderm-like cells, which secrete plasminogen activator (38) of the parietal endoderm type (Strickland, S., personal communication) and produce small amounts of type IV collagen and laminin but no alpha-fetoprotein (39,40). This behavior is consistent with parietal endoderm. These H6 endoderm-like cells also secrete a factor - as yet unidentified - which stimulates the survival and growth of undifferentiated H6 stem cells in low serum-containing medium (41), analogous to that reported for the PC13 teratocarcinoma line (42,43). Sublines resistant to 6-thioguanine, 5-bromodeoxyuridine, ouabain, chloramphenicol, or retinoic acid have been derived, and these used, for example, to show that terminally differentiating H6 cells can be rescued by hybridization to undifferentiated stem cells (38).

H6 cells are usually grown in minimal essential medium (MEM) which lacks calcium (SMEM) but is supplemented with 5% fetal calf serum, so that the final calcium concentration is approximately 0.1 mM. In this medium H6

Table 1: Plating Efficiency of H6 ECC in Increasing Calcium Concentrations

Expt. No.	Cell Type	$CaCl_2$ Concentration		
		2mM	4mM	7mM
I	B1	16 %	49 %	67 %
	6B	15 %	22 %	47 %
II	B1	25.5%	53.5%	70 %
	6B	48 %	50.5%	62.5%

100 cells of the H6 ECC sublines, B1 or 6B, were inoculated into 25 cm^2 gelatin-coated tissue culture petri dishes in MEM with 10% fetal calf serum, plus added $CaCl_2$ when necessary to reach the concentrations indicated. After 5 days the colonies were fixed, stained, and counted. Each plating efficiency is the average of duplicate dishes. The B1 colonies in medium containing 7 mM $CaCl_2$ were relatively small.

Table 2: Aggregation and Compaction of H6 cells in Increasing Calcium Concentrations

Calcium Concentration	Total No. of Cells/Dish	% of Maximum	Morphology of Aggregates
0.1mM	6.0×10^6	87	Loosely aggregated
2 mM	6.9×10^6	100	Variably compacted
3 mM	6.4×10^6	93	Well compacted
4 mM	5.0×10^6	72	Well compacted
5 mM	4.2×10^6	61	Well compacted
6 mM	3.5×10^6	51	Well compacted
7 mM	3.2×10^6	46	Well compacted

4×10^5 H6 cells were inoculated into 25 cm^2 bacteriological petri dishes containing fetal calf serum-supplemented low calcium MEM, MEM, or MEM plus added $CaCl_2$ to give the concentrations indicated. The cells were immediately agitated by rotation at 70 rpm in a humidified incubator at 37°C to encourage aggregation. After 48 hours the contents of each dish was transferred to a 75 cm^2 gelatin-coated tissue culture petri dish, with added medium calculated to produce a final $CaCl_2$ concentration close to 2 mM to encourage attachment. After stationary incubation for a few hours, the cells were trypsinized and counted in a Coulter counter. The values represent the average of duplicate dishes. Two parallel stationary monolayer cultures in MEM containing 2 mM $CaCl_2$ averaged a total of 8.6×10^6 cells, that is, 125% of the maximum growth of the rotated culture in 2 mM $CaCl_2$ medium.

cells grow loosely attached to the cell culture flask as single cells or pairs of cells, dividing about every 16 hours. Attachment can be encouraged by coating the surface with gelatin and the use of medium containing more calcium, such as MEM itself with 5% serum, which contains about 2 mM calcium. In fact, higher concentrations of calcium lead to increased attachment and colony formation (Table 1), although at the highest concentration tested (7 mM) some inhibition of colony growth occurs.

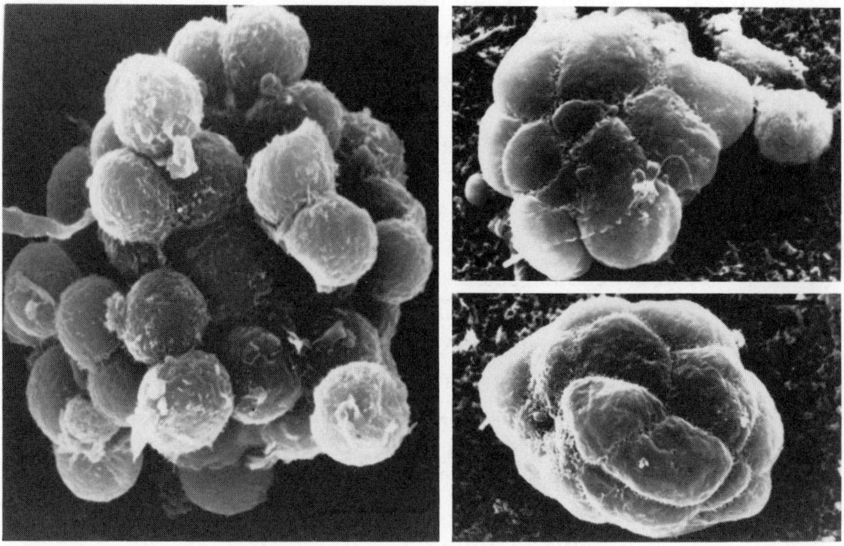

Figure 1: Scanning electron micrographs of uncompacted (left) and compacted (right) aggregates of embryonal carcinoma cells. H6 cells were stirred in suspension for 48 hours in culture medium containing 0.5 mM (left) or 4 mM (right) calcium. The resulting aggregates were collected by brief centrifugation (100 × G), immediately fixed in 2.5% glutaraldehyde in 0.1 M phosphate buffer (pH 7.2), and then prepared for microscopy by critical point drying, with the help of Dr. Alan Walker. The uncompacted and compacted aggregates are shown at the same magnification, with the bars representing 10 microns.

If H6 cells are grown in suspension at a concentration of calcium (0.5 mM) five times higher than that used in routine maintenance and are constantly agitated to discourage attachment, the cells adhere to one another, forming loose, grape-like aggregates (Figure 1) (44). Three variant sublines which aggregate poorly have been derived. Each showed a diminished response to peanut agglutinin, and in one line this has been traced to a partial deficiency of peanut agglutinin receptor (45,46). If the calcium concentration is raised to twenty times that used in maintenance (2 mM), as in MEM, the aggregates undergo compaction within an hour (Figure 1) (Table 2) (40,46). The importance of calcium in compaction is well known (8,47,48); but most other ECC lines are grown already compacted in monolayer culture, so the H6 line is unusual in that the steps of aggregation and compaction can be separated by the use of different calcium concentrations.

In H6 cells, as in other ECC lines, the process of compaction has many similarities to that which occurs in the morula, such as inhibition by cytoskeletal poisons (including tetracaine, colcemid, vinblastine, colchicine, and cytochalasin B), by interference at surface sites by exposure to lectins (such as wheat germ agglutinin or concanavalin A), or by interruption of glycosylation by exposure to tunicamycin or reaction with anti-H6 Fab fragments or anti-F9 (46). In both the morula and H6 cell aggregates, compaction is reversed by removal of calcium.

As occurs in the uncompacted early 8-cell stage embryo, small areas of gap junctions are present in uncompacted H6 aggregates (49), and the extent of these increases with compaction. We recently used three different measures of cell-to-cell coupling to show that some crossfeeding occurs through the small clusters of gap junctions in loose aggregates, while more occurs through the larger clusters in compacted aggregates (50). These three techniques included survival in HAT medium of aggregates containing a mixture of 6-thioguanine- and 5-bromodeoxyuridine-resistant sublines of H6 due to mutual crossfeeding, killing of 6-thioguanine-resistant cells in mixed aggregates by crossfeeding of 6-thioguanine derivatives from 6-thioguanine-sensitive cells, and transfer of ^{3}H-thymidine derivatives into 5-bromodeoxyuridine-resistant cells from 5-bromodeoxyuridine-sensitive cells. These studies suggested that the enlarged areas of gap junctions in compacted aggregates represented the growth of existing junctions or the addition of new junctions, rather than simply the coalescence of those pre-existing in the uncompacted aggregates.

As in the morula (5,12,15), signals transmitted through gap junctions in ECC aggregates probably do not initiate compaction and are at most permissive. The known growth factors are too large in size to pass through gap junctions. Cell-to-cell coupling may simply serve to promote nutrition by the transfer of small cellular constituents and metabolites unable to cross the cell membrane, facilitating the maintenance of a similar intracellular microenvironment in adjacent cells. However, the number or function of gap junctions in the morula and ECC aggregates between external cells, between internal cells, or between both, might differ, especially as the differentiation of external cells proceeds. It has been reported that monolayer cultures of the PC13 ECC line show metabolic cooperation with endodermal cells taken from the early embryo, as well as with mesoderm and embryonic ectoderm, but not with trophectoderm nor its derivatives (51). Such "communication compartments" are known to develop in post-implantation mouse embryos, partially segregating trophectoderm, embryonic ectoderm and extra-embryonic endoderm (52,53). Unfortunately, we have been unable to isolate a metabolic cooperation-deficient subline of H6 ECC (54), which might give additional information on the role of gap junctions in ECC compaction, but the use of inhibitors of metabolic cooperation may be helpful (55,56).

Considerable amounts of extracellular matrix materials are laid down just below the external cell layer of PSA1 embryoid bodies and of F9 aggregates exposed to low concentrations of retinoic acid. When H6 cell aggregates (or monolayers) are exposed to retinoic acid, with or without cyclic AMP, the production of these compounds is also increased, but only minimally in comparison to F9 cells (40). Also, an external layer of endoderm is not formed in H6 aggregates, and alpha-fetoprotein is not produced. Unlike H6 cells, F9 cells are usually grown in monolayer culture in a compacted state, and this might facilitate the subsequent differentiation of aggregates of these cells. Or it may be that F9 cells are already partially differentiated, and this is enhanced by retinoic acid. However, it is most likely that H6 cells and F9 cells are inherently (genetically) different in regard to the ability to differentiate, as occurs for other attributes of ECC lines (24).

In any case, because the stages of aggregation and compaction can be clearly separated in H6 cells, they provide an opportunity to examine in detail the process of compaction. Toward this goal we recently isolated two variant H6 sublines which are unable to compact as aggregates (57). First, the method of producing aggregates was changed from cell-to-cell adhesion in agitated suspension culture to growth and division of individual cells in medium containing 1.3% methylcellulose, so that each aggregate was derived clonally from a single original cell. Then variant clones unable to compact were isolated directly by identification and picking under the microscope. There were no spontaneously occurring compaction-defective variants, so the parental cultures first had to be exposed to the mutagen, nitrosoguanidine. Prior to our work, a compaction-defective subline had been isolated by a less direct method from a mutagenized F9 culture (58). This has subsequently been shown to contain about one-half the usual amount per cell of the surface glycoprotein uvomorulin (Adamson, E. D., personal communication), which plays an important role in cell adhesion and compaction (59,60). Upon testing our two compaction-defective variants, we have now found one to completely lack uvomorulin, even when exposed to added calcium, which increases the amount of uvomorulin in wild-type cells (61). Hybrid cells in which these variant cells were fused to wild-type H6 cells also compacted poorly and lacked uvomorulin. This result, in addition to the complete absence of uvomorulin in the variant itself, suggests that the mode of action of the mutation is dominant, perhaps involving an interruption of the usual induction of uvomorulin by calcium. Our other H6 variant contained uvomorulin in the usual amount and with the usual electrophoretic mobility. Using methylcellulose-containing medium in the same way, it might be possible to directly isolate differentiation-deficient variants among F9 cell aggregates treated with retinoic acid (but not among H6 aggregates, which do not form embryoid bodies, as noted above). Alternatively, rare undifferentiated F9 aggregates might be isolated by density gradient centrifugation, fluorescence-activated cell sorting, or some sort of immunological, nutritional, or drug selection which favors aggregates lacking an external layer of visceral endoderm. Such variants might not be uncommon, as a differentiation-defective variant has already been isolated from the PSA1 line by screening the morphology of a relatively low number of clonal colonies (62).

Because aggregates of H6 cells compact well but differentiate poorly, in contrast to F9 cell aggregates, we recently developed a subline of F9 cells which grows uncompacted in monolayer culture in low calcium medium, like H6 cells (40). When these cells were trypsinized and cultured in suspension in the same low calcium medium, they formed loose, open aggregates, again like H6 cells. These aggregates failed to form embryoid bodies after exposure to retinoic acid, confirming the requirement of compaction for differentiation. If, after culture for three days, the uncompacted F9 aggregates were exposed to added calcium, all compacted within an hour, again like H6 cells. The number of days for aggregates to acquire this ability to compact rapidly was reduced if the monolayer of cells from which the aggregates were derived had been exposed to added calcium to cause compaction for several days *prior* to trypsinization and aggregation. Next, treatment of the compacted F9 aggregates with retinoic acid was followed by differentiation into embryoid bodies. The number of days required for this also was reduced if the aggregates were formed from previously compacted cells, presumably because compaction of the aggregates was completed sooner. These studies suggest that when a monolayer of uncompacted F9 cells is exposed to added calcium, certain proteins important for compaction are synthesized; and some of these persist through trypsinization to be carried over from monolayer to aggregates. It may be possible to test this hypothesis with inhibitors of protein synthesis. An intriguing alternative hypothesis is that an inhibitor of compaction is present in uncompacted F9 cells, which gradually decreases in the presence of added calcium (16). Indeed it seemed possible that the dominant H6 mutant described above might constitutively overproduce this same inhibitor of compaction! However, we were not able to demonstrate such an inhibitor either in medium conditioned by the mutant cells nor in mixed aggregates containing mutant and wild-type cells (Littlefield, J.W., unpublished results).

Information about the timing of transcription of messages and the synthesis of proteins essential for compaction and differentiation in the early mouse embryo has recently become available (7,16,63). Changes in the transcription of various genes early and late during the retinoic acid-induced differentiation of monolayer cultures of ECC, such as F9, have also been reported (64-70). Similar information about aggregates of ECC (or ES cells) which are differentiating an external endoderm layer would be interesting as well, since this situation more closely resembles that in the embryo. To start in this direction, we have made a cDNA library of the genes transcribed two hours after the compaction of H6 aggregates in response to added calcium. This library is now being screened with cDNA probes made from the mRNA of aggregates before and after compaction in order to search for messages which are increased or decreased following compaction. In the future we want to extend this work to the F9 system described above, in which cells in stepwise fashion aggregate and compact, like H6 cells, but then can differentiate as well. Libraries will be obtained at multiple time points from before compaction to full differentiation. We hope to compare the differentiating cells in the external layer with the undifferentiated internal cells; this might be possible by the uptake of iron particles into the external cells or tagging of the external cells with antibodies specific for endoderm markers, followed by decompaction and separation of the labelled and unlabelled cells.

In general, at the present stage of understanding of early mammalian embryogenesis, and with still limited techniques to study the small amounts of material available, it appears that ECC and ES lines, and particularly aggregates of these, may be able to provide information relevant to compaction and differentiation in the embryo, as well as to the formation of epithelia in general. The accessibility of such systems to manipulation makes it likely that existing theories concerning the relation between compaction and differentiation can be tested rigorously. Additional theories may arise. For example, the internal cells of a compacted aggregate (or morula) may simply be constrained physically by their neighbors, and this constraint may somehow maintain an undifferentiated (or totipotent) state. Commonly in cell culture the central cells of colonies growing on a monolayer are small, multi-layered, and actively dividing, while the peripheral cells are large, spread out, and perhaps differentiated (38). This might be due to the number of divisions which external cells have undergone, but it seems equally plausible that some cells need space to enlarge as they differentiate, and without this they may remain undifferentiated.

Finally, there is more and more reason to believe that various growth factors and their receptors function in early embryogenesis (43,71,72), but presently it would be technically difficult to establish the roles of these factors unequivocally. The possible physiological functions of factors which stimulate or inhibit growth or differentiation can be examined more easily in ECC (and ES cell) monolayers (43,71,73-77), and perhaps in aggregates. For example, as mentioned above, differentiated PC13 and H6 cells secrete a factor which stimulates their undifferentiated parental stem cells to grow (41-43). In aggregates of ECC such as the PSA1 and F9 lines, and in ES cell aggregates, the external layer of endoderm-like cells may secrete a similar factor, which stimulates the growth of the internal cells and perhaps maintains their pluripotentiality. Possibly this same factor brings about the burst of division which occurs in the embryonic ectodermal cells of the mouse embryo at the egg cylinder stage (6-7 1/2 days), when they are surrounded by visceral endoderm. Exploration of the role of growth and differentiation factors in early embryogenesis may proceed most profitably through further studies with ECC and ES lines, and then with mouse embryos cultured *in vitro*, before utilizing mouse embryos *in vivo*.

ACKNOWLEDGEMENT

Research by the author mentioned in this article was supported by NCI grant number CA 16754.

REFERENCES

1. Martin, G. R. (1980). *Science*, **209**:768-776.
2. Hogan, B. L. M., Barlow, D. P., and Tilly, R. (1983). *Cancer Surveys* **2**:115-140.
3. Johnson, M. H., and Maro, B. (1986). **In**: *Experimental Approaches to Mammalian Embryonic Development* (Rossant, J., and Pedersen, R.A., Eds.) pp. 35-65, Cambridge University Press, New York, **4**:459-485.
4. Fleming, T. P., and Johnson, M. H. (1988). *Annu. Rev. Cell Biol.*, **4**:459-485.
5. Goodall, H., and Johnson, M. H. (1982). *Nature*, **295**:524-526.
6. Goodall, H., and Johnson, M. H. (1984). *J. Embryol. Exp. Morph.*, **72**:53-76.
7. McLachlin, J. R., and Kidder, G. M. (1986). *Devel. Biol.*, **117**:146-155.
8. Ducibella, T., and Anderson, E. (1975). *Devel. Biol.*, **47**:45-58.
9. Pratt, H. P. M. (1985). *J. Embryol. Exp. Morph.*, **90**:101-121.
10. Fleming, T. P., Cannon, P. M., and Pickering, S. J. (1986). *Devel. Biol.*, **113**:406-419.
11. Johnson, M. H., Ziomek, C. A., Reeve, W. J. D., Pratt, H. P. M., Goodall, H., and Handyside, A. H. (1984). **In**: *Ultrastructure of Reproduction* (Van Blerkom, J., and Motta, P.M., Eds.) pp. 205-217, Martinus Nijhoff Publishers, Boston.
12. Goodall, H. (1986). *J. Embryol. Exp. Morph.*, **91**:283-296.
13. Mintz, B. (1964). *J. Exp. Zoology*, **157**:85-100.
14. Tarkowski, A. K., and Wroblewska, J. (1967). *J. Embryol. Exp. Morph.*, **18:155-180.**
15. Johnson, M. H., and Ziomek, C. A. (1981). *Cell*, **24**:71-80.
16. Levy, J. B., Johnson, M. H., Goodall, H., and Maro, B. (1986). *J. Embryol. Exp. Morph.*, **95**:213-237.

17. Stevens, L. C. (1958). *J. Nat. Cancer Inst.*, **20**:1257-1275.
18. Pierce, G. B., and Dixon, F. J. (1959). *Cancer*, **12**:584-589.
19. Evans, M. J., and Kaufman, M. H. (1981). *Nature*, **292**:154-156.
20. Martin, G. R. (1981). *Proc. Natl. Acad. Sci. USA*, **78**:7634-7638.
21. Axelrod, H. R. (1984). *Devel. Biol.*, **101**:225-228.
22. Doetschman, T. C., Eistetter, H., Katz, M., Schmidt, W., and Kemler, R. (1985). *J. Embryol. Exp. Morp.*, **87**:27-45.
23. Bradley, A., Evans, M., Kaufman, M. H., and Robertson, E. (1984). *Nature*, **309**:255-256.
24. Papaioannou, V. E., and Rossant, J. (1983). *Cancer Surveys*, **2**:165-183.
25. Martin, G. R., and Evans, M. J. (1975). *Proc. Natl. Acad. Sci. USA*, **72:1441-1445.**
26. Martin, G. R., and Evans, M. J. (1975). *Cell*, **6**:467-474.
27. Martin, G. R. (1977). *Devel. Biol.*, **61**:69-83.
28. Hogan, B. L. M., Taylor, A., and Adamson, E. (1981). *Nature*, **291**:235-237.
29. Strickland, S., Smith, K. K., and Marotti, K. R. (1980). *Cell*, **21**:347-355.
30. Grover, A., Oshima, R. G., and Adamson, E. (1983). *J. Cell Biol.*, **96**:1690-1696.
31. Grover, A., Andrews, G., and Adamson, E. D. (1983). *J. Cell Biol.*, **97**:137-144.
32. Grover, A., and Adamson, E. D. (1985). *J. Biol. Chem.*, **260**:12252-12258.
33. Grover, A., Edwards, S. A., Bourdon, M., and Adamson, E. D. (1987). *Differentiation*, **36**:138-144.
34. Grover, A., and Adamson, E. D. (1986). *Devel. Biol.*, **114**:492-503.
35. Grover, A., and Adamson, E. D. (1986). *In Vitro Cellular and Developmental Biology*, **22**:280-284.
36. Hay, E. D. (1984). **In**: *The Role of Extracellular Matrix in Development* (Trelstad, R. L. Ed.), pp. 1-31, Alan R. Liss, Inc., New York.
37. Reddi, A. H. (1984). **In**: *Extracellular Matrix Biochemistry* (Piez, K. A., and Reddi, A. H., Eds.), pp. 375-412, Elsevier, New York.
38. Littlefield, J. W., and Felix, J. S. (1982). *Somatic Cell Genetics*, **8**:743-757.
39. Felix, J. S., and Ostrand-Rosenberg, S. (1983). *Am. J. Human Gen.*, **35**:163A.
40. Littlefield, J. W. (1989*b*). *Devel. Genetics*, in press.
41. Germain, E. L., and Littlefield, J. W. (1986). *In Vitro Cellular and Developmental Biology*, **22**:107-112.
42. Isacke, C. M., and Deller, M. J. (1983). *J. Cell Physiol.*, **17**:407-414.
43. Heath, J. K., and Rees, A. R. (1986). **In**: *Growth Factors in Biology and Medicine* (Ciba Foundation Symposium 116) pp. 3-22, Pitman Publishing, Ltd, London.
44. Felix, J. S., Kartha, S., and Littlefield, J. W. (1982). **In**: *Membranes and Genetic Disease* (Sheppard, J. R., Anderson, V. E., and Eaton, J. W., Eds.) pp. 173-192, Alan R. Liss, Inc., New York.
45. Kartha, S., Felix, J. S., and Littlefield, J. W. (1985). *Differentiation*, **29**:77-81.
46. Kartha, S., Felix, J. S., and Littlefield, J. W. (1983). *Devel. Biol.*, **100**:172-180.
47. Hyafil, L. F., Babinet, C., and Jacob, F. (1981). *Cell*, **26**:447-454.
48. Bilozur, M. and Powers, D. R. (1982). *Exper. Cell Res.*, **142**:39-45.
49. Strum, J. M., Kartha, S., and Felix, J. S. (1984). *Devel. Biol.*, **105**:93-101.
50. Littlefield, J. W. (1987). *Devel. Biol.*, **121**:288-291.
51. Gaunt, S. J. (1979). *J. Embryol. Exp. Morph.*, **54**:263-275.
52. Lo, C. W., and Gilula, N. B. (1979). *Cell*, **18**:399-409.
53. Caveney, S. (1985). *Annu. Rev. Physiol.*, **47**:319-335.
54. Hooper, M. L., and Subak-Sharpe, J. H. (1981). *Internat. Rev. Cytol.*, **69**:45-105.
55. Yotti, L. P., Chang, C. C., and Trosko, J. E. (1979). *Science*, **206**:1089-1091.
56. Smith, T. A., Smith, A. G., and Hooper, M. L. (1986). *Exper. Cell Res.*, **165**:417-428.
57. Littlefield, J. W. (1989*a*). *Devel. Genetics*, in press.
58. Grover, A., Rosenstraus, M. J., Sterman, B., Snook, M. E., and Adamson, E. D. (1987). *Devel. Biol.*, **120**:1-11.

59. Ringwald, M., Schuh, R., Vestweber, D., Eistetter, H., Lottspeich, F., Engel, J., Dolz, R., Jahnig, F., Epplen, J., Mayer, S., Muller, C., and Kemler, R. (1987). *EMBO J.*, **6**:3647-3653.
60. Takeichi, M. (1988). *Development*, **102**:639-655.
61. Littlefield, J. W., and Whitehouse, L. L. (1989). In preparation.
62 Felix, J. S. (1985). *Differentiation*, **29**:254-258.
63. Kidder, G. M., and McLachlin, J. R. (1985). *Devel. Biol.*, **112**:265-275.
64. Wang, S-Y., and Gudas, L. J. (1983). *Proc. Natl. Acad. Sci. USA*, **80**:5880-5884.
65. Young, P. R., and Tilghman, S. M. (1984). *Mol. Cell. Biol.*, **4**:898-907.
66. Levine, R. A., LaRosa, G. J., and Gudas, L. J. (1984). *Mol. Cell. Biol.*, **4**:2142-2150.
67. Wang, S-Y., LaRosa, G. J., and Gudas, L. J. (1985). *Devel. Biol.*, **107**:75-86.
68. Griep, A. E., and DeLuca, H. F. (1986). *Proc. Natl. Acad. Sci. USA*, **83**:5539-5543.
69. Edwards, S. A., Rundell, A. Y. K., and Adamson, E. D. (1988). *Devel. Biol.*, **129**:91-102.
70. LaRosa, G. J., and Gudas, L. J. (1988). *Mol. Cell. Biol.*, **8**:3906-3917.
71. Adamson, E. D. (1987). **In**: *Experimental Approaches to Mammalian Embryonic Development* (Rossant, J., and Pedersen, R. A., Eds.), pp. 1-51, Cambridge University Press, New York.
72. Mercola, M., and Stiles, C. D. (1988). *Development*, **102**:451-460.
73. Heath, J. K. (1983). *Cancer Surveys*, **2**:141-164.
74. Nagarajan, L., Anderson, W. B., Nissley, P. S., Rechler, M. M., and Jetten, A. M. (1985). *J. Cell Physiol.*, **124**:199-206.
75. Rizzino, A., and Bowen-Pope, D. F. (1985). *Devel. Biol.*, **110**:15-22.
76. Jakobovits, A., Banda, M. J., and Martin, G. R. (1985). **In**: *Cancer Cells 3 / Growth Factors and Transformation* (Feramisco, J., Ozanne, B., and Stiles, C., Eds.) pp. 393-399, Cold Spring Harbor Laboratory, New York.
77. Rizzino, A., Kuszynski, C., Ruff, E., and Tiesman, J. (1988). *Devel. Biol.*, **129**:61-71.

3

MELANINS IN THE PHOTOBIOLOGY OF SKIN CANCER AND THE RADIOBIOLOGY OF MELANOMAS

Helene Z. Hill

Section of Cancer Biology
Department of Radiology
New Jersey Medical School
Newark, NJ 07103-2757

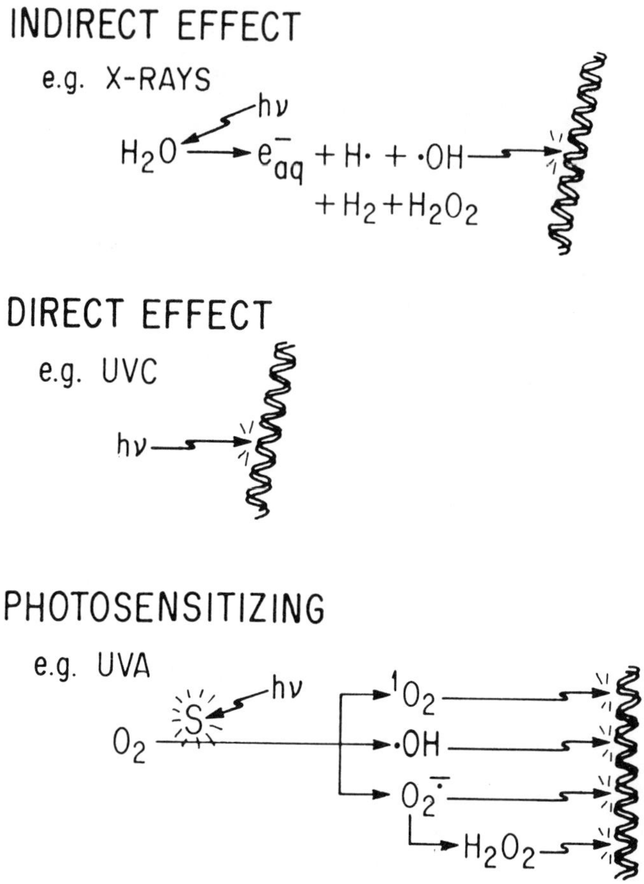

Figure 1: Radiation damage to DNA. Indirect Effect: Photons of low LET ionizing radiations interact with matter, mainly water in biological systems, to produce ionizations which ultimately result in the formation of free radicals, the most important of which is the hydroxyl radical. This, in turn, reacts with biological molecules to produce damage. The target for lethal, mutagenic and carcinogenic damage is almost certainly DNA. Direct Effect: Photons of UVC and UVB interact directly with biological molecules to produce damage. Photosensitizing: Photons of UVB and UVA activate an intracellular photosensitizer such as a flavin or a porphyrin which causes the formation of active oxygen species which can in turn react with target molecules such as DNA.

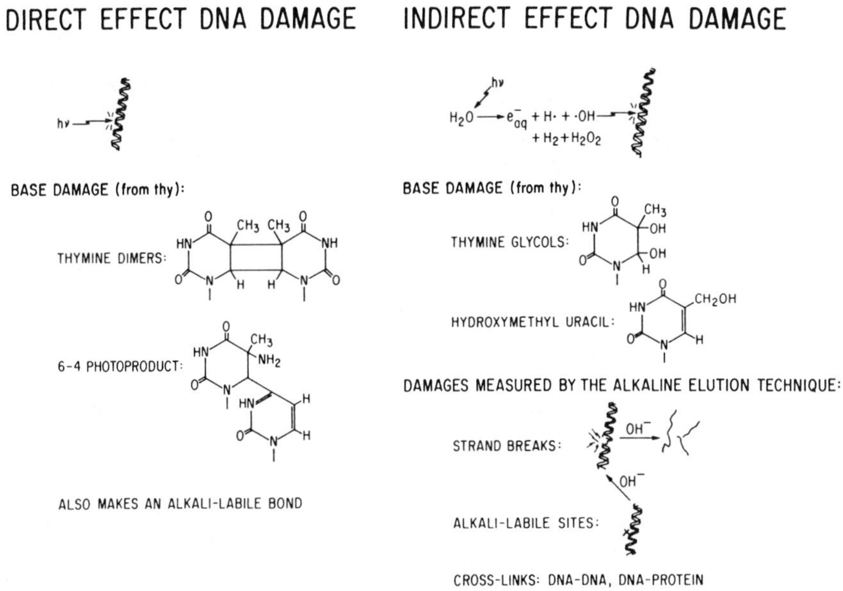

Figure 2: Direct and indirect effect DNA damages. All four DNA bases can be damaged by radiations, but most is known about thymine interactions since the precursor dThd in radioactive form incorporates directly into DNA without further metabolism. Thymine glycols, hydroxymethyl uracil, DNA strand breaks, and DNA protein crosslinks are indirect effect damages that have been found in cellular DNA. Base damages such as apurinic, apyrimidinic and 6-4 photoproduct produce strand breaks after incubation with alkali. Direct effect damages observed in cellular DNA include cyclobutane pyrimidine dimers and the 6-4 photoproduct.

Melanomas are resistant to ionizing radiation therapy and black-skinned individuals rarely get skin cancer. Although the radiations involved are part of the same energy continuum, low LET ionizing radiations act by causing primarily indirect effect lesions in DNA, while solar energies probably cause both indirect (by photosensitizing) effect and direct effect damage (Figures 1 and 2). A major lesion caused by the direct effect of solar ultraviolet light (UV) is the pyrimidine dimer (py<>py) which has been shown to be mutagenic and carcinogenic (1,2). Perhaps equally important is the so-called 6-4 photoproduct (3). Indirect effect solar lesions include DNA strand breaks and DNA-protein crosslinks (4-7).

The common denominator in radiation resistance of melanomas and dark skin is the pigment, melanin. If melanin has a role in radiation resistance of melanoma, it must act as a radical scavenger, since low LET ionizing radiation lethality is caused by free radical interaction with DNA (indirect effect).

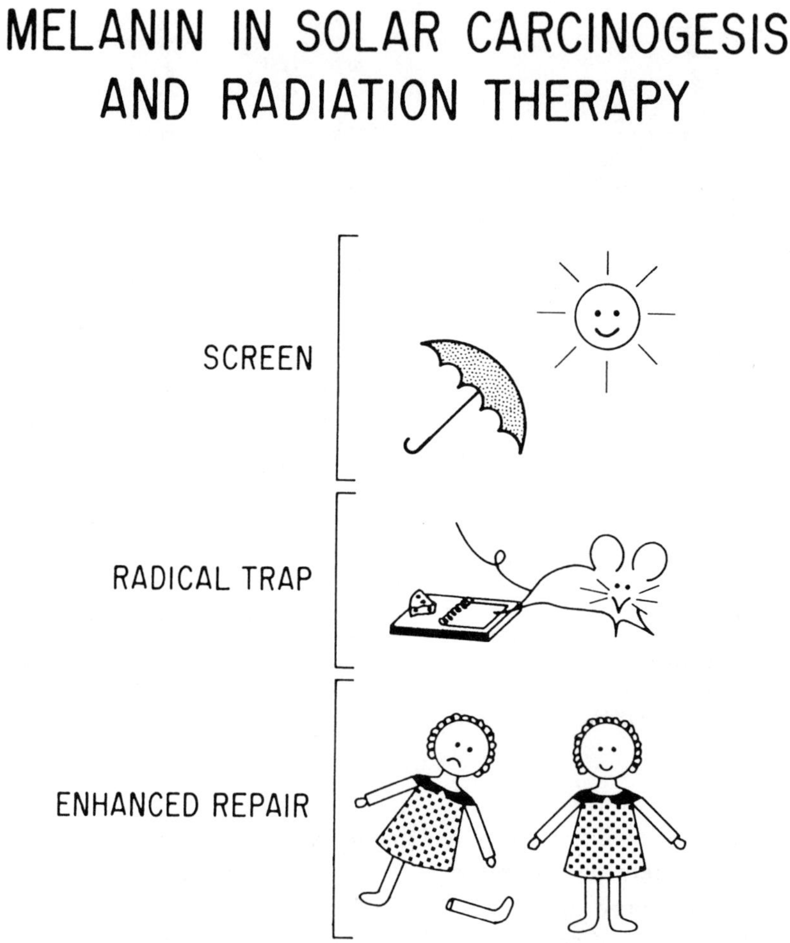

Figure 3: Three possible mechanisms for melanin as a protector against photon damage.

Melanin might act by directly scavenging radicals, but this is an unlikely mechanism which would require its presence in the nucleus in close proximity to the DNA. Alternatively, it might reduce the overall radical balance in the cells by acting as a free radical sink, as, for example, by making the cells functionally hypoxic. The cartoons in Figure 3 suggest 3 possible mechanisms

Table 1: Estimated Incidence of and Mortality from Skin Cancer in the US for 1987			
Type	**Incidence**	**Deaths**	**Deaths/Incidence %**
Basal cell epitheliomas	400,000	460	0.12
Squamous cell carcinomas	100,000	1,540	1.5
Malignant melanomas	25,800	5,800	22.5

of action for melanin. Perhaps some aspects of all three are important in its effectiveness.

Skin cancer frequencies in the U.S. are shown in Table 1. The ratio of non-melanomas in whites versus blacks is 68:1 (8), while for melanoma it is 9:1 (9). Melanoma is, in fact, the most important skin cancer in blacks. Keratinocytes give rise to the non-melanoma skin cancers: basal cell cancers arising in the germinal layer, and squamous cell cancers which come from the outer layers. There are many more keratinocytes than there are melanocytes, the cells that give rise to melanomas, and the keratinocytes divide much more frequently. Thus, the melanocytes are less numerous and quiescent, yet they give rise to more tumors than would be predicted on the basis of their numbers and their cell division frequencies. Sun tanning and burning stimulate cell division in the skin, but proportionately more keratinocytes than melanocytes are so stimulated (10).

RADIATION AND MELANOMA

In the first radiation survival curves, melanomas appeared to have broad shoulder regions, reflecting resistance, but normal slopes (11,12). Later, Hill *et al.* (13) showed that B16 melanoma tumors and clonal lines generally had normal shoulders, but slopes indicative of relative radiation resistance. Rofstad *et al.* (14) in Norway studied the ionizing radiation responses of 38 human melanomas and 5 clones from a single nude mouse xenograft (Figure 4). The survival parameters of the melanomas were compared to *in vitro* survivals of 3 clinically radiation sensitive and 3 clinically radiation resistant non-melanoma tumor cells. They conclude that the melanomas tend to exhibit a broad spectrum with respect to intrinsic sensitivity to radiation. They can be as sensitive as the most sensitive tumors and they can be as resistant as the most resistant tumors. There is no general rule except that they are probably much more heterogeneous than any other type of tumor, additionally emphasized by the responses of the clones derived from the same tumor.

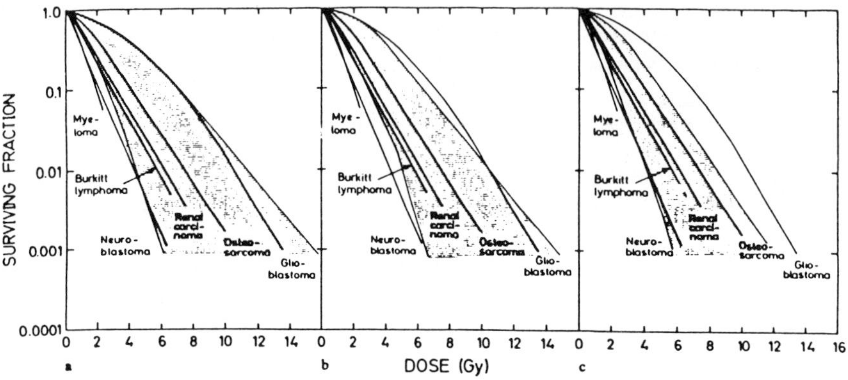

Figure 4: The survival characteristics of human melanomas (shaded areas) were compared to three clinically radiation sensitive tumors (myeloma, neuroblastoma, Burkitt lymphoma) and three clinically resistant tumors (renal carcinoma, gliblastoma and osteosarcoma) using three different types of assay: established cell lines (a), disaggregated xenografts (b) and disaggregated surgical specimens (c) (from Rofstad, reference 14, reprinted with permission from E. K. Rofstad).

Melanomas also tend to be hypoxic. To study this phenomenon, human melanomas are grown in nude mice and irradiated. Their subsequent behavior is quantitated and compared to other tumors. One melanoma described by Guichard *et al.* (15) had an hypoxic fraction that was estimated to be between 85 and 100%. In a comprehensive review (16) 79 hypoxic fraction determinations were made for rodent tumors. The average hypoxic fraction was 18 ± percent.

An example of the effect of hypoxia on radiation response is shown in Figure 5 using B16 mouse melanoma (17). Tumor-bearing mice were irradiated, the tumor cells were removed, plated in tissue culture and the colonies were quantitated as a function of radiation dose after culture. In the upper curve, the mice were irradiated under anoxic conditions, that is, they were killed and then irradiated 5 minutes later, by which time all of the oxygen would have been consumed. The lower curve (air-breathing) breaks and shows a tail that parallels the upper (hypoxic) curve. These are the resistant hypoxic cells in the tumor, the fraction of which can be estimated by curve-fitting. In this case, it is approximately 12%.

Unfortunately, melanin is rarely quantitated in such studies. There is some evidence, however, that melanin and radiation resistance are related. Rofstad *et al.* (18) compared the radiation response of two human melanomas in nude

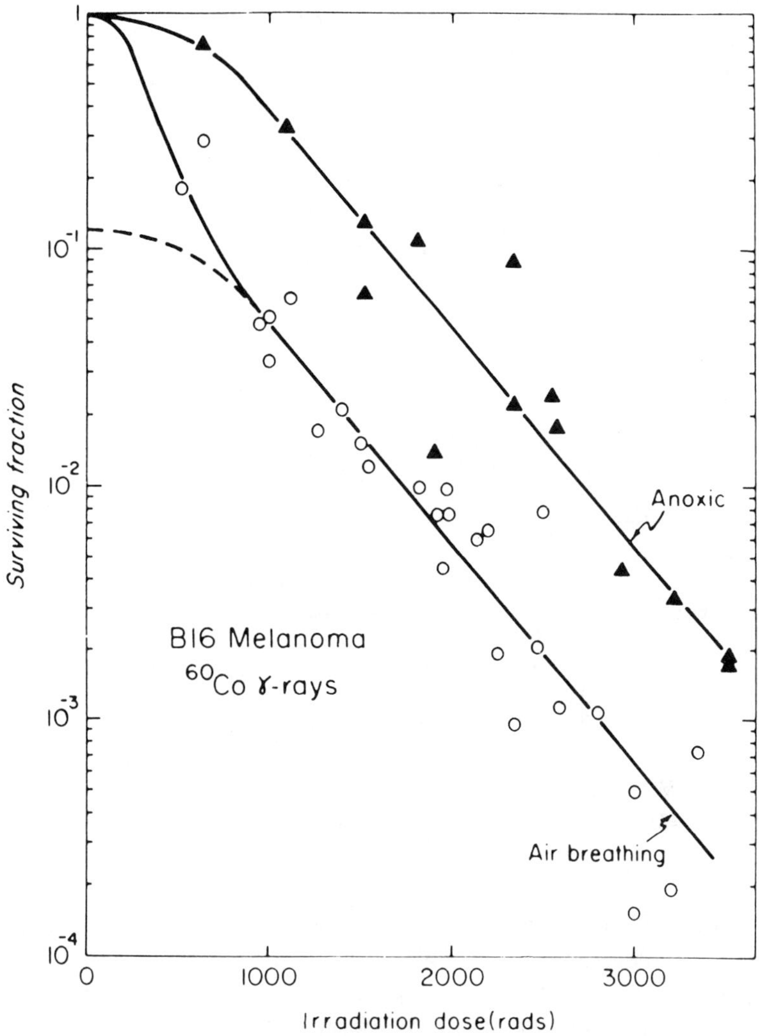

Figure 5: Demonstration of an hypoxic fraction in B16 melanomas growing in mice. In air-breathing animals, the initial slope represents the sensitivity of the well-oxygenated tumor cells, the final slope, which is parallel to the curve for anoxic cells, shows the sensitivity of the hypoxic cells in the tumors (from Hill and Stanley, reference 17, reprinted with permission from R. J. Hill ©Waverly Press, Inc., 1975; dashed line has been added).

mice. The black tumor was not controlled by a fractionation scheme that controlled the pale one. Cobb (19) separated pigmented and unpigmented regions of melanoma in mice and found that tissue slices of the former were

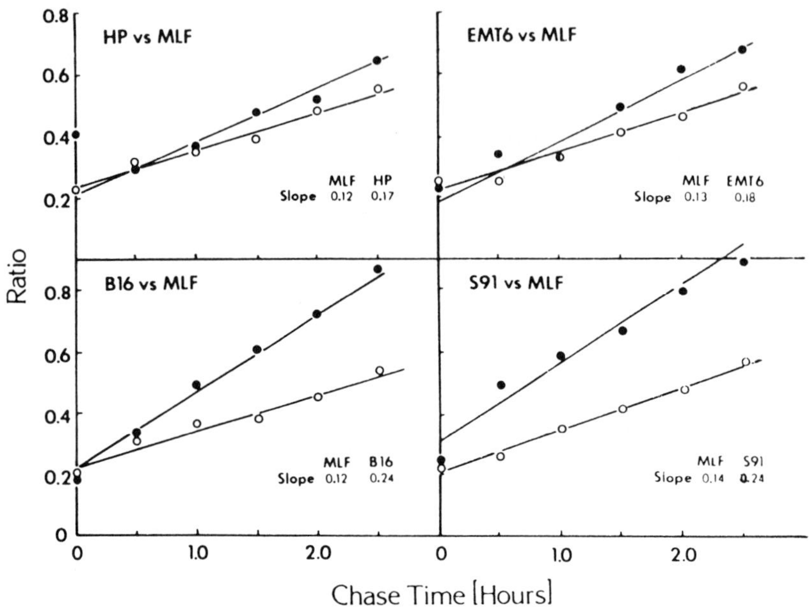

Figure 6: The rate of post-replication repair in three mouse melanoma lines compared to that of mouse lung fibroblasts. B16 and S91 cells have a rate that is twice that of MLF. HP shows a similar rate to MLF, but when this tumor line was first isolated, the rate was also twice that of MLF. EMT6 is a mouse mammary carcinoma line. Its rate is between the melanomas and MLF (from Hill and Setlow, reference 20, ©Waverly Press, Inc., 1980).

less sensitive to radiation than those from the latter. In these 2 cases, however, hypoxia could be responsible for the relative resistance of the more pigmented tumors.

We have found that a very melanotic variant of murine Cloudman S91 melanoma is quite resistant to ionizing radiation lethality compared to an otherwise isogenic amelanotic variant, but, in this case, differences in repair have not been ruled out (unpublished observations).

DNA REPAIR AND MELANOMA

Hill and Setlow (20) measured post-replication repair in several mouse melanoma lines compared to that in a non-melanoma line, irradiated (with UVC) at the same time (Figure 6). Several human melanomas have also been studied, one of which was from an amelanotic tumor. All of these had increased post-replication repair relative to a non-melanoma kidney tumor line (21).

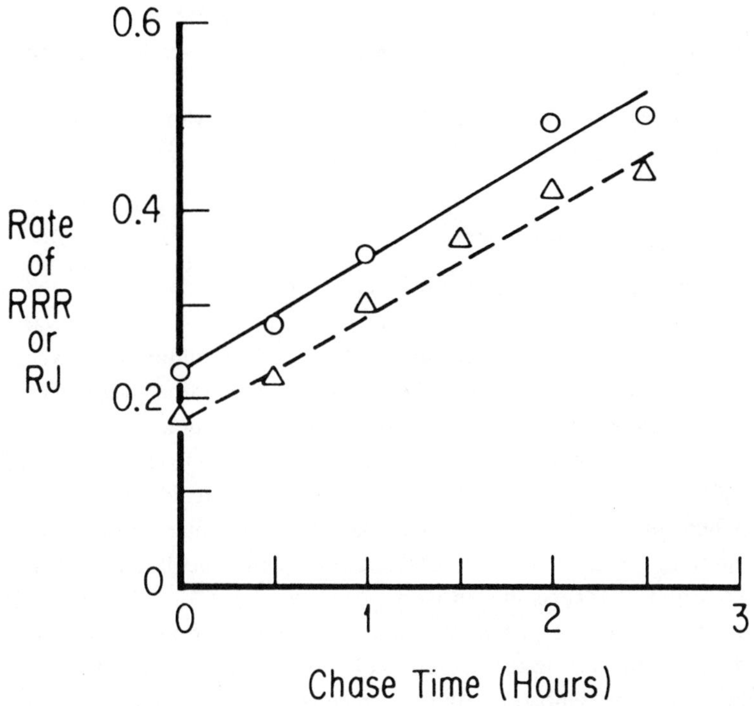

Figure 7: The rate of post-replication repair (triangles) after 10 J/m^2 of UVC in a human melanoma cell line (MeWo) compared to the rate of replicon joining (circles, no UV) in the same cells. The two rates are the same, although the length of newly synthesized DNA pieces is slightly smaller after UV, as shown by the Y-intercepts of the two curves.

Post-replication repair is measured by determining the rate of joining of small pieces of newly synthesized DNA that result after a short period of replication using an endogenous damaged template. The damage is usually inflicted by irradiating the cells with UVC which induces the formation of py<>py in the DNA. Post-replication repair is actually a misnomer. It is a process which indicates that some cells can replicate their DNA at a near-normal rate in spite of the fact that they contain lesions. We found one human melanoma

line in which the rate of joining newly synthesized DNA in the presence of dimers was the same as the rate of replicon joining in the DNA in the absence of dimers (Figure 7).

The increased capacity of melanomas to synthesized DNA on damaged templates may reflect an ability these cells have to replicate DNA in the face of constant bombardment by damaging radicals, for as we shall see, melanin, itself, may be very genotoxic.

Melanomas, then, tend to have large hypoxic fractions, may be intrinsically resistant to ionizing radiation, and have increased rates of post-replication repair.

MELANOCYTES AND SKIN

In skin, the germinal or basal layer gives rise to the cells in the outer layers. These cells continue to replicate as they progress upward towards the surface of the skin. The penultimate layer is the granular layer where the nuclei are extruded, giving rise to the dead, horny layer, the stratum corneum, which contains no live cells, no nuclei, and little melanin. The optics of this arrangement are such that considerable light is reflected from the skin surface, especially in darker skinned individuals. Melanocytes reside in the suprabasal layer where their dendrites, neuronal-like processes, interdigitate among the keratinocytes. Melanocytes synthesize melanin in subcellular bodies called melanosomes. Mature melanosomes migrate out the dendrites to be phagocytized by the keratinocytes. Since the transfer of melanosomes occurs near the basal layer, and the suprabasal keratinocytes continue to divide and distribute fewer melanosomes into more cells, there is a gradient of melanin in the skin with the highest concentration deepest in the skin. The light-scattering nature of melanin at the surface may offer adequate solar protection against skin cancer in dark-skinned peoples, but orientals are also quite free of skin cancer, yet their melanin content is low. A number of skin reactions to sunlight correlate better with the minimal erythema dose than with pigmentation (22) within a given skin type, which is to say that the reddening response to sunlight does not necessarily correlate inversely with degree of pigmentation. Here lies one of the paradoxes of melanin.

MELANINS

Figure 8 shows schematically the still somewhat hypothetical biosynthetic pathway for melanins. Only one enzyme is thought to be involved. The first precursor is tyrosine, which is converted by tyrosinase to DOPA. This is further converted *via* a number of undefined or ill-defined intermediates to DOPA quinone. It is believed that the major precursor of melanin is indole-5,6-quinone.

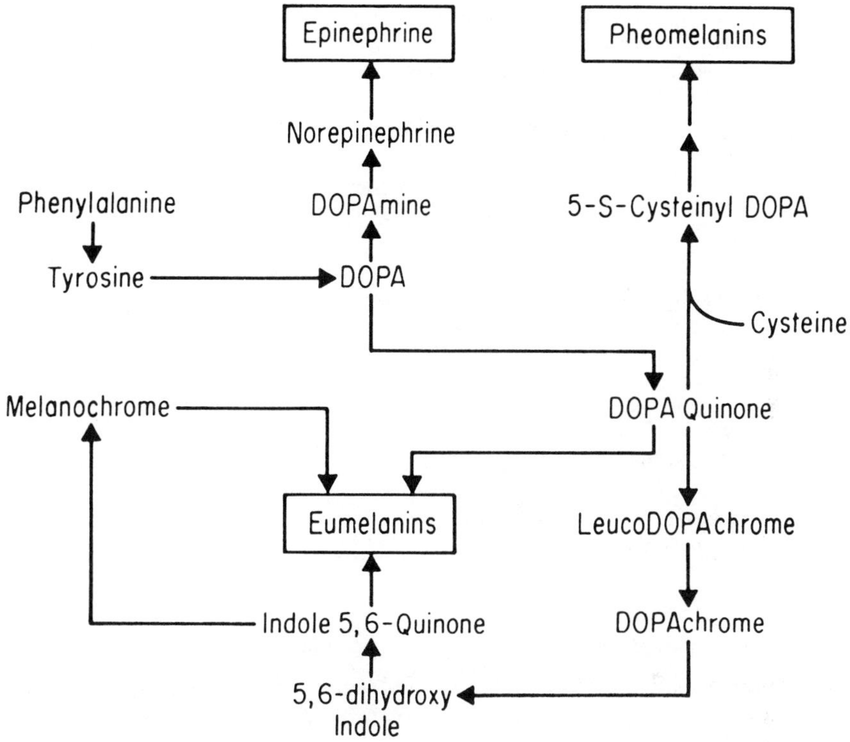

Figure 8: The biosynthesis of melanins and related compounds. The details of melanin synthesis are not completely worked out. Only one enzyme, tyrosinase, appears to be required, and many of the steps occur spontaneously.

The biosynthetic pathway leading to the formation of melanin has steps in common with biosynthesis of epinephrine, an important neurotransmitter. Both nerve cells and melanocytes are derived embryonically from ectodermal neural crest. Furthermore, although melanins are found in many body tissues, including heart and liver, their major locations are skin and nervous tissue. Even more interesting is the observation that the two types of tumor that tend to be most resistant to ionizing radiation are melanomas and brain tumors (glioblastomas). Both these tumors are quite rare in blacks. The functions of neural melanins are not known, but at least one writer believes that these pigments are the ultimate organizing molecules, mediating all functions of nerves from impulse transmission to control of emotions (23).

Table 2: Biological Properties Of Melanins
Major pigment for camouflage
Present in some organisms of all biological kingdoms
High inverse relationship between skin pigmentation and solar carcinogenesis
Present primarily in skin and nervous tissue
Found in many other organs
Melanomas, tumors of melanocytes, are often resistant to ionizing radiation

Table 3: Physical and Chemical Properties of Melanins
Amorphous
Insoluble
Irregular polymer of indoles, quinones, phenols, pyrroles derived from DOPA
Polymerizes spontaneously or enzymatically from DOPA
Emits continuous free-radical signal in the absence of stimulation
In the presence of light
scavenges radicals: superoxide, singlet oxygen, hydroxyl radical
consumes oxygen
enhances endogenous free radical signal
emits heat, radio signals, hydrogen peroxide, superoxide
hydroxyl radical can form by a Fenton-type reaction
Binds drugs such as chlorpromazine.

There are actually two classes of animal melanins: eumelanins, brown and black pigments, and phaeomelanins, red and yellow pigments. They are synthesized in a similar manner, as shown in Figure 8, but in the synthesis of phaeomelanin, DOPA quinone reacts with cysteine to form 5-S-cysteinyl DOPA which is thought to be the major precursor. Pheomelanin is the melanin in red hair and possibly in freckles. It is more photo-reactive and light labile than eumelanin, and emits toxic radicals on irradiation (24). It can also damage DNA (25). Individuals with the phenotype of red hair and freckles are very skin cancer-prone. Eumelanins extracted from biological tissues are amorphous and insoluble, hence difficult to analyze chemically. Phaeomelanins are more soluble.

Probably all natural melanins are mixtures of pheo- and eumelanins. For the purposes of the ensuing discussion, melanin(s) will refer to eumelanin(s) unless stated otherwise.

Table 2 summarizes the major biological properties of melanins. In addition to the properties already discussed that are associated with melanins, they are the major pigment for camouflage, and are present in some organisms of all biological kingdoms. The pigment in the sustantia nigra of the brain is neural melanin. It is absent from this area in Parkinson's Disease. Recently, in California, a number of patients were recognized as having drug-induced Parkinsonism. The drug involved, MPTP, was a contaminant in a bad batch of heroin. MPTP binds melanins (26). At post mortem, victims were found to have amelanotic substantias nigra.

The chemical, physical and photochemical properties of melanins are summarized in Table 3. As noted, melanins are amorphous insoluble polymers when extracted from biological tissues. Soluble eumelanins can be obtained by *in vitro* synthesis either by auto-oxidation of DOPA or by enzymatic synthesis. The synthetic product has many characteristics in common with the cellular pigments, but any extrapolations must be made with caution.

A unique property of melanins is the continuous emission of free radical (Electron spin resonance, ESR) signals without any stimulation (27). ESR signals can be induced in other biological macromolecules but emission without stimulation is very rare in the biological world. The ESR signal can be enhanced by radiation (28,20).

The optical density of melanin increases as the wave length of light decreases. All melanins are similar, in this respect, although they may have some minor differences. The spectrum is rather flat through the visible but increases considerably at about 300 nanometers (29-31).

When melanins are stimulated by light, they take up oxygen, emit superoxide and they darken (28,32). After long, or intense, exposure they bleach. The phenomenon known as immediate pigment darkening in sun-exposed skin is due to solar-induced darkening of epidermal melanin. Melanin synthesis is not involved, although sunlight induces melanin synthesis and transfer of melanosomes to keratinocytes several hours after sun exposure. Melanins have been shown to effectively scavenge the radicals produced in water by ionizing radiation (33). They have the highest affinity for the hydroxyl radical which is thought to be the most important radical biologically.

Melanin is dark and absorbs light. It also is an active free-radical scavenger. A reasonable working hypothesis to unify the radiation resistance of melanomas and the solar resistance of black skin would seen to be that it protects DNA from direct effect damage by screening, and that it also protects DNA from indirect effect damage by radical scavenging.

The preliminary studies reported in the literature reflect some ambiguities, in this regard.

In 1970, workers in Germany (34) measured the ionizing radiation survival of mouse L-cells with respect to colony-forming ability. One aliquot of cells was irradiated in the presence of melanin (conditions unstated) and the other in its absence. The reported results indicate that melanin (source unstated) was radioprotective.

Table 4A: Relative Effectiveness of Compounds at Competing for Thy- and dThd-Damaging Radicals

Compound	ED50			
	dThd		Thy	
	monomer (mM)	mg/ml	monomer (mM)	mg/ml
Trp	0.11 ± 0	0.02		
Tyr	0.12 ± 0.006	0.02		
DOPA	0.12 ± 0.004	0.02		
His	0.14 ± 0.001	0.02		
phe	0.16 ± 0.006	0.03		
dThd	0.17 ± 0.003	0.04		
thy			0.19 ± 0.007	0.02
cys	0.20 ± 0.006	0.02	0.40 ± 0.003	0.05
S-eumelanin	1.6 ± 0.12	0.24	1.0 ± 0.04	0.15
DNA	2.4 ± 0.35	0.82		
BSA	2.5 ± 0.08	0.30		
Ficoll	4.3 ± 0.14	0.77	7.5 ± 0.14	1.35
ala	4.3	0.38		
gly	4.3	0.32		

The mg/ml column shows the actual concentration of a given substance at the ED50 (the concentration of radical scavenger that inhibits radiolysis of the target molecule by 50%). The lower the ED50, the more effective the compound. For polymers, the monomer MW was calculated from that of the known monomer, e.g. glucose for ficoll, a sucrose polymer, or by averaging known monomer MWs, as for DNA and albumin. For eumelanins, the monomer was assumed to be indole-5,6-quinone, and for phaeomelanin, it was assumed to be 5-S-cysteinyl DOPA (from Hill *et al.*, reference 37). S-eumelanin is persulfate-initiated auto-oxidized DOPA-melanin from the Sigma Chemical Company, St. Louis, MO.

Two years later a group in the U.S. (35) looked at the effect of a solar lamp (spectral qualities undefined) on macrophages that were incubated with and without melanin for about 4 hours and irradiated. The cells irradiated in the presence of melanin were more sensitive to the lethal effects of the lamp than the controls. This result is entirely contrary to the first result.

Plant and fungal melanins lack nitrogen but have many properties in common with animal melanins. Two melanin pigments were extracted from a fungus and were incubated with penicillium cells which were then assayed for survival after UV and after ionizing radiation. The pigments were called 3A and 3B. 3A was very protective against UV, while 3B has almost no UV protective effect. However, 3B was more protective against ionizing radiation than 3A (36). These findings serve to demonstrate that individual melanin pigments may have very dissimilar photobiological, radiobiological, photochemical and radiochemical properties.

Table 4B: Relative Effectivness of Melanins at Competing for dThd Destroying Radicals at Comparable Optical Densities. Sepia MFA is a small MW (2000 Dalton) fraction obtained by gentle oxidation of melanin from ink sacs. It was the gift of the Clairol Corp., Stamford, CT. DM-melanin is melanin obtained by autoxidation of DOPA.

Melanin	**%dThd Remaining Mean ± s.d.**	**Melanin concentration**	
		mg/ml	**Monomer (mM)**
None	30 ± 3		
S-eumelanin	59 ± 3	0.270	1.7
Sepia MFA	69 ± 8	0.100	0.67
DM-melanin	53 ± 4	0.087	0.58
Phaeomelanin	57 ± 7	0.077	0.31

THE RADICAL SCAVENGING PROPERTIES OF MELANINS COMPARED TO OTHER BIOLOGICAL MACROMOLECULES

In order to assess the effectiveness of melanins as radical scavengers relative to other biological macromolecules, solutions of thymine and thymidine were exposed to ionizing radiation in the presence of a variety of solutes. The amount of target molecule that survived was quantitated by HPLC. Table 4 shows that the best protection — or competition for radicals — was seen with small molecules, but that melanins were the most effective macromolecules (37).

Table 4 also compares the scavenging properties of several melanins. These were all used at the same optical density, based on the assumption that optical density reflects concentration of functional groups. Clearly, this is the case, since protective abilities, on this basis, are quite similar, although actual concentrations are very different. The best radical scavenger appears to be phaeomelanin, followed by Sepia MFA, a small molecular weight eumelanin derived from squid ink. Phaeomelanin is also the most photochemically and photobiologically active of the melanins.

From these studies, it is concluded that melanins are fairly good radical scavengers: not as effective as small molecules, but better than other cellular macromolecules. Small molecules such as tyrosine, phenylalanine, DOPA and cysteine are not present in cells in high concentrations because they are toxic. The cell may take advantage of the useful properties of such small molecules by polymerizing them into melanins to reduce their toxicity.

CELLULAR MELANIN AND DNA DAMAGE

As noted, one of the major photoproducts from direct effect UV damage to DNA is the pyrimidine dimer. Such dimers can be quantitated by isolating and purifying UV-damaged DNA, incubating it with an enzyme fraction from

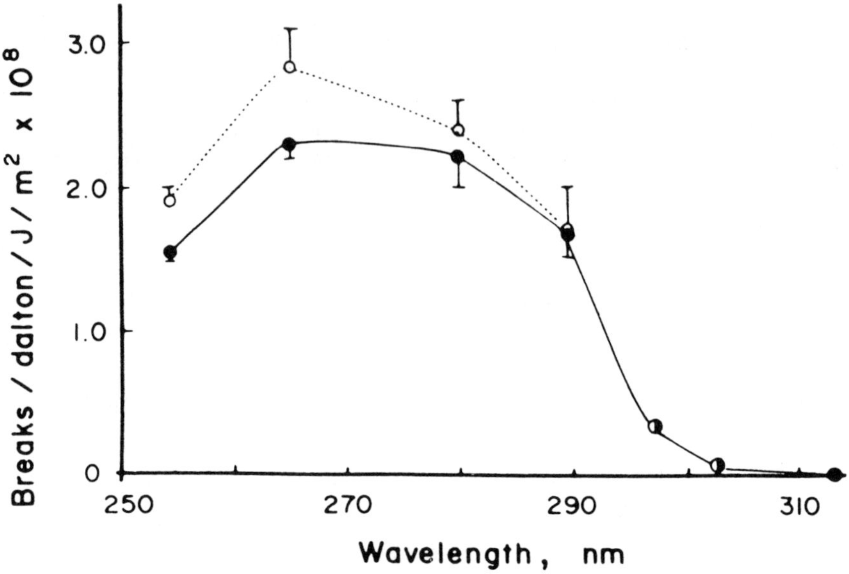

Figure 9: Action spectra for UV-endonuclease sensitive sites in DNA of EMT6 (open circles) and Cloudman S91 melanoma cells (closed circles). There are more dimers produced in the DNA of EMT6 cells at wave lengths less than 290 nm. At this wave length and above, the dimer production in the two cell lines is the same, indicating that melanin in the S91 cells does not protect against the production of these lesions at environmentally important wave lengths (from Hill and Setlow, reference 38, ©Pergamon Press Ltd., 1982).

Micrococcus luteus with UV-endonuclease activity. This enzyme recognizes pypy in DNA by causing a single strand break at the dimer site. The MW of such treated DNA can be compared to untreated or undamaged DNA in order to quantitate dimers. The method is sensitive and quite specific. Hill and Setlow (38) compared dimers in melanotic Cloudman S91 cells to EMT6, a non-melanoma mammary carcinoma cell line at various wave lengths. The results are shown in Figure 9. It would seen that the melanin in the melanoma cells was protective against dimer induction at wave lengths below, but not above 280 nm. Solar wave lengths that penetrate to the earth's surface, have a cut off of 290 nm. Therefore, the protective ability of melanin in these cells would appear to be unimportant in any biological sense. These findings could be the result of too low a concentration of melanin in the cells, or melanin in a form that does not protect at higher wave lengths and, therefore, different from dermal melanin in dark skin. These findings could also mean that in solar exposed melanocytes and keratinocytes, py<>py are not lesions that can lead to cancer, since there may be no difference in the induction of these lesions in white and black skin at solar wave lengths.

If melanin in the Cloudman S91 cells is radioprotective, then melanotic cells should be resistant to radiation relative to amelanotic cells. To test this, we compared the survival of colony-forming ability of otherwise isogenic melanotic and amelanotic S91 lines. The melanotic line was about 30% *less* sensitive to ionizing radiation but it was *more* sensitive to UVC and even more sensitive to UVB. UVB and ionizing radiation both involve reactions of DNA with active oxygen species. These results clearly demonstrate that different active species are involved at these two different energy ranges, and that the cells handle the damages in different fashions.

In other studies we have radio-labelled the DNA of the melanotic and the amelanotic S91 lines and of a mammary carcinoma line and exposed them to high doses of UVC. DNA was isolated from these cells, and from unirradiated cells, degraded to nitrogen bases by acid hydrolysis and the bases were separated by HPLC. Some interesting findings have emerged (38a).

After UVC, there was less thymine dimer in the DNA of the melanotic cells compared to the other two cell lines. However, new thymine lesions were apparent in the DNA from the melanotic cells. One of these lesions has been identified as thymine glycol — a lesion present in DNA after ionizing radiation. The other lesions probably arose by further oxidation of thymine glycol during acid hydrolysis of the DNA. We concluded that the primary lesion is thymine glycol.

Further study indicated that the amount of thymine glycol found in the DNA depended both on the UVC dose and the amount of melanin in the cells. The carcinoma cells, lacking melanin, had no thymine glycol, while the amelanotic cells which contain a small amount of melanin had less than the melanotic cells. These findings imply that melanin protects against direct effect DNA damage but in some way leads to the production of indirect effect DNA damage. However, the doses of UV administered were beyond the biologically significant range. At lower doses, such indirect effect damage might be obviated by other cellular protective mechanisms. The fact that the melanotic cells are more sensitive to killing by UVC suggests that the oxidative lesions in the melanotic cells do indeed produce biologically relevant damage. The greater sensitivity of the melanotic cells to UVB killing further implies that there are disproportionately more oxidative lesions in DNA of melanotic cells at longer wave lengths.

UVA is thought to damage DNA by photosensitization: photons interact with light-absorbing molecules in the cell to produce active oxygen species which in turn damage DNA. DNA-protein crosslinks are believed to arise in this manner. Melanotic S91 cells and EMT6 mammary carcinoma cells were compared for the UVA induction of DNA-protein crosslinks (39). There were fewer such lesions in the melanotic cells. Apparently melanin can quench the photosensitization reaction induced by UVA. Curiously, the induction of DNA-protein crosslinks by UVC was the same in the two cell lines, although the carcinoma cells have fewer pyrimidine dimers than the melanotic cells after UVC. Here is another paradox raised by melanin!

EXOGENOUS MELANIN AND DNA DAMAGE

Pajak *et al.* (40) studied the effect on certain cellular functions of incubation of exogenous melanins with Chinese hamster ovary cells. These cells would only take up melanin in particulate form by phagocytizing it. Once this occurred, the cells consumed oxygen rapidly, but the ingested melanin did not protect the cells against the lethality of ionizing radiation (41). Since the radical species — hydroxyl radical — that is thought to be most important in ionizing radiation lethality is very short lived, if melanin acts directly as a radioprotector it would be most likely to do so only if it could penetrate to the nucleus. This would be more likely if the melanin were soluble. In our experiments, then, we used soluble melanin and B16CL4 melanoma cells. The melanoma cells were used in the hope of circumventing the phagocytosis problem which the Chinese hamster ovary cells presented. The melanin used was synthesized by autooxidation of D,L-DOPA. Similar melanin preparations had been used to study ESR signals, oxygen uptake and superoxide production.

The original experimental plan was to study the effect of this melanin in protection against ionizing radiation induction of DNA strand breaks measured by alkaline elution analysis (42), but it was seen that the melanin, itself, was DNA damaging (43). DNA damage occurs quickly and is maximal at between 15 and 20 minutes. After this time, even though the melanin continues to be present in the medium, the breaks are repaired. If the cells are washed and reincubated in the absence of melanin, the breaks are repaired more rapidly. Therefore, melanin either inhibits repair or continues to damage the DNA, but at progressively slower rates. The extent of strand breaks and their disappearance in the presence of melanin is shown in Figure 10.

The concave nature of the elution profiles could be due to the induction by melanin of DNA cross-links. That this is not likely is shown by irradiating the cells after incubation with melanin and comparing that elution pattern with the elution after ionizing radiation alone. If the melanin were inducing crosslinks, it should retard the elution of the irradiated DNA. However, in this situation, the breaks caused by the melanin appeared to be additive with the

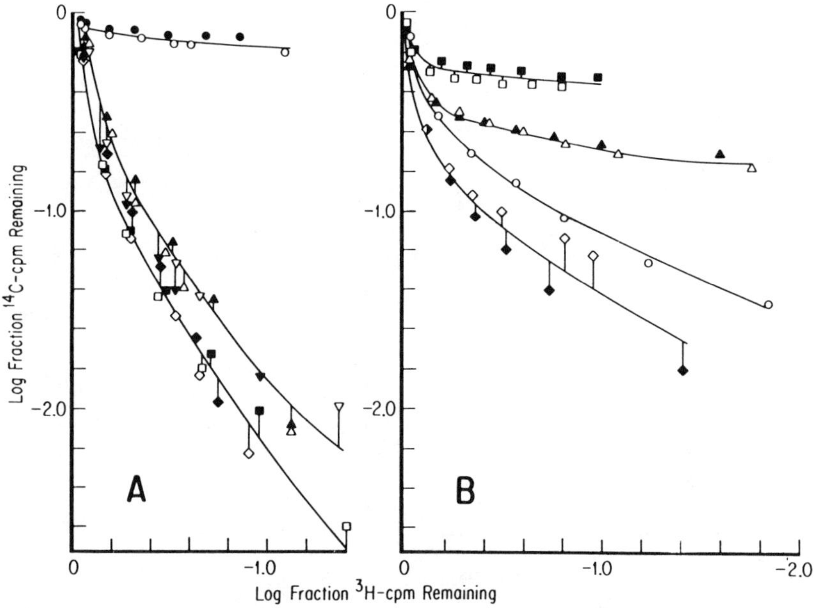

Figure 10: DNA strand breaks caused by melanin. Eumelanin from auto-oxidation of D,L-DOPA causes strand breaks in DNA in a time-dependent manner. A. Few breaks appear if the cells are rinsed with melanin (circles). Maximum breaks occur at 15 to 20 minutes (squares and diamonds). The effect is submaximal at 5 and 10 minutes (triangles). B. Continued incubation in the presence of melanin shows that the breaks can be repaired: diamonds 30 minutes, circles 1 hour, triangles 1.5 hours, squares 3.5 hours (from Hill and Hill, reference 43, ©Alan R. Liss, Inc., 1987).

breaks caused by ionizing radiation. Therefore, the elution rate after the combined treated was greater than after either treatment alone.

Melanins from other sources also proved to be DNA damaging in this system. These are shown in Figure 11. Some melanins were inactive at causing strand breaks directly, but, when the cellular DNA from the lysed cells was kept at high pH overnight, alkali-labile damage in the DNA was also revealed.

Cells that have been incubated with melanin for 15 minutes, washed and assayed for colony-forming-ability, show that melanin is toxic in the same concentration range that causes strand breaks in the DNA.

From these experiments we conclude that melanin can damage DNA, the damage is repairable, even in the continued presence of melanin in the medium, and the damage is additive with ionizing radiation damage. There does not seem to be any indication that this melanin is radioprotective.

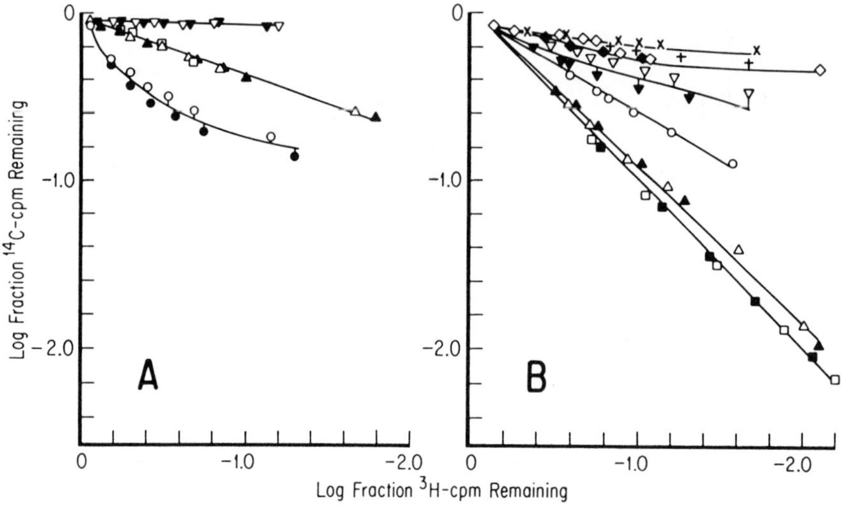

Figure 11: DNA strand breaks and alkali-labile sites caused by several melanins. Melanins were incubated with cells for 30 minutes at 0.5 mg/ml. D,L-DOPA-melanin (circles); Sepia melanin (squares); Sigma melanin (triangles); Sepia melanin-free acid (MFA), a small (2000) MW melanin fraction derived from Sepia melanin (inverted triangles); L-DOPA-melanin (diamonds); O melanin (+'s, x's). A. Frank strand breaks: L-DOPA-melanin and O melanin treated cell DNA had the same elution profile as MFA. B. Alkali-labile sites (from Hill and Hill, reference 43, ©Alan R. Liss, Inc., 1987).

SUMMARY

Melanins are photochemically and photobiologically active substances. They can themselves be DNA damaging, probably by indirect means by producing free radicals, and they can protect DNA from damage. In producing whatever biological effect, their chemical and physical nature must be very important to their function.

Figure 12 summarizes the current thinking in this laboratory regarding the way that melanins work in 3 situations: in black skin where they prevent sunburn and cancer, in white skin where they do not do these things and in melanomas where they enhance radioresistance.

In black skin, the melanosomes are large, round, single and densely pigmented. Here, they act as a screen. Light cannot penetrate to the nucleus, so free radicals and direct interactions cannot occur. Whether melanins in this form can themselves produce DNA damage will be determined in the future.

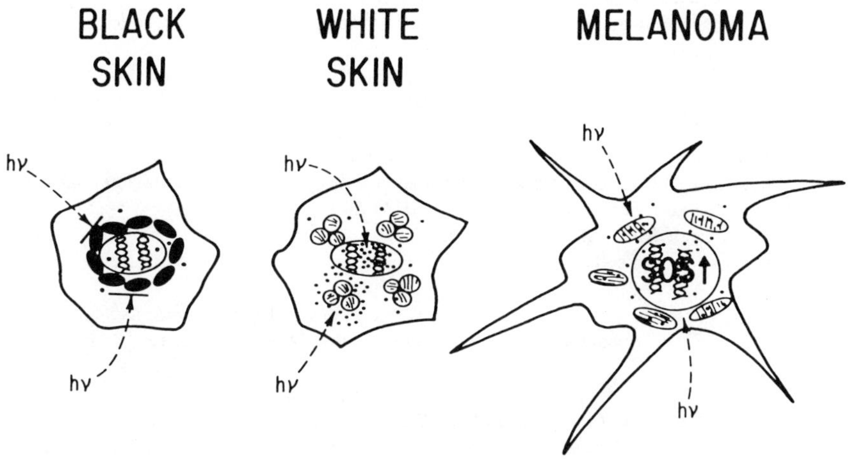

Figure 12: Hypothetical models to explain the behavior of melanins in different cells. In black skin, dense melanosomes effectively shield the nucleus from deleterious light, but produce an occasional potentially damaging free-radical. In white skin, the aggregated melanosomes are poor shields, permitting light to penetrate to the nucleus. Furthermore, interaction of the loosely packed melanin with photons results in the production of free-radicals. In melanoma, the same phenomena may occur, but here, in some, but not all tumors, DNA repair is enhanced, as denoted by the SOS distress call in the nucleus.

If they do, then DNA repair should also be enhanced in such cells. That they may is indicated by the greater relative frequency of melanomas in blacks.

In white skin, the melanosomes are small, oval, aggregated and lightly pigmented. They are poor sun screens. Pigment may more easily escape to cause DNA damage. The situation may be worse when the skin is induced to tan. There may be, in fact, no such thing as a "safe tan". In very pale individuals, the melanin precursors are present in greater concentration due to an inability to tan and these are more genotoxic than the polymeric melanin.

In melanoma, melanin is presumed to act as a radical scavenger, making the cells functionally hypoxic. It also may induce DNA repair. Since melanomas are heterogeneous in their radiation response, melanins in different tumors must not be equally effective.

REFERENCES

1. Smith, K. C. (1977). Ultraviolet radiation effects on molecules and cells, **In**: *The Science of Photobiology*, K. C. Smith, Ed. Plenum Press, New York. pp. 113-142.
2. Hart, R. W. Setlow, R. B., and Woodhead, A. D. (1977). Evidence that pyrimidine dimers in DNA can give rise to tumors. *Proc. Natl. Acad. Sci. USA*, **74**:5574-5578.

3. Mitchell, D. L., and Nairn, R. S. (1989). The biology of the (6-4) photoproduct. *Photochem. Photobiol.*, **49**:805-819.
4. Peak, J. G., Peak, M. J., and MacCoss, M. (1987). DNA breakage caused by 334-nm ultraviolet light is enhanced by naturally occurring nucleic acid components and nucleotide coenzymes. *Photochem. Photobiol.*, **39**:713-716.
5. Peak, J. G., Peak M. J., and Foote, C. S. (1986). Observations on the photosensitized breakage of DNA by 2-thiouracil and 334-nm ultraviolet radiation. *Photochem. Photobiol.*, **44**:111-116.
6. Peak, M. J., Peak, J. G., and Jones, C. A. (1985). Different (direct and indirect) mechanisms for the induction of DNA-protein crosslinks in human cells by far- and near-ultraviolet radiations (290 and 405 nm). *Photochem. Photobiol.*, **42**:141-146.
7. Peak, J. G., Peak, M. J., Sikorski, R. S., and Jones, C. A. (1985). Induction of DNA-protein crosslinks in human cells by ultraviolet and visible radiations: action spectrum. *Photochem. Photobiol.*, **41**:295-302.
8. Scotto, J., Fears, T. R., and Fraumeni, J. R., Jr., (1982). **In**: *Cancer Epidemiology and Prevention*, D. Schottenfeld and J. F. Fraumeni, Jr., Ids. (W.B. Saunders Co., Philadelphia), pp. 254-276.
9. Young, J. L., Jr., and Polack, E. S. (1982). **In**: *Cancer Epidemiology and Prevention*, D. Schottenfeld, and J. F. Fraumeni, Jr., Eds. (W.B. Saunders Co., Philadelphia), pp. 138-165.
10. Potten, C. S. (1985). *Radiation and Skin*, Taylor and Francis, Philadelphia, pp. 187-203.
11. Dewey, D. L. (1971). The radiosensitivity of melanoma cells in culture. *Brit. J. Radiol.*, **44**:816-817.
12. Barranco, S. C., Romsdahl, M. M., and Humphrey R. M. (1971). The radiation response of human malignant cells grown *in vitro*. *Cancer Res.*, **31**:830-833.
13. Hill, H. Z., Hill, G. J., Miller, C. F., Kwong, F., and Purdy, J. (1979). Radiation and melanoma: Response of B16 mouse tumor cells and clonal lines to *in vitro* irradiation. *Radiation Research*, **80**:259-276.
14. Rofstad, E. K. (1986). Radiation biology of malignant melanoma. *Acta Radiologica*, **25**:1-9.
15. Guichard, M., Gosse, C., and Malaise, E-P. (1977). Survival curve of human melanoma in nude mice. *J. Natl. Cancer Inst.*, **58**:1665-1669.
16. Moulder, J. E., and Rockwell, S. (1984). Hypoxic fractions of solid tumors: Experimental techniques, methods of analysis, and a survey of existing data. *Int. J. Radiat. Oncol. Biol. Phys.*, **10**:695-712.
17. Hill, R. P., and Stanley J. A. (1975). The response of hypoxic B16 melanoma cells to *in vivo* treatment with chemotherapeutic agents. *Cancer Res.*, **35**:1147-1153.
18. Rofstad, E. K., Brustad, T., and Johannessen, J. V. (1977). Effect of clinically relevant irradiation regimes on human malignant melanomas grown in athymic nude mice. *Acta Radiol. Ther. Phys. Biol.*, **16**:273-280.
19. Cobb, J. P. (1956). Effect of *in vitro* X irradiation on pigmented and pale slices of Cloudman S91 mouse melanoma as measured by subsequent proliferation *in vivo*. *J. Natl. Cancer Inst.*, **17**:657-666.
20. Hill, H. Z., and Setlow, R. B. (1980). Postreplication repair in three murine melanomas, a mammary carcinoma, and a normal mouse lung fibroblast line. *Cancer Res.*, **40**:1867-1872.
21. Hill, H. Z., Hill, G. J., and Kurowski, S. M. (1984). Post replication repair after UV in human amelanotic and melanotic melanomas. *Proc. Amer. Assoc. Cancer Res.*, **25**:108.
22. Kolias, N., Baqer, A., and Sadiq, I. (1987). The absorption characteristics of human "melanin" in the ultraviolet. *Photochem. Photobiol.*, **45**(suppl.):975.
23. Barr, F. R., Saloma, J. S., and Buchele, M. J. (1983). Melanin: the organizing molecule. *Med. Hypoth.*, **11**:1-140.
24. Agin, P. P. Sayre, R. M., and Chedekel, M. R. Photodegradation of phaeomelanin: an *in vitro* model. *Photochem. Photobiol.*, **31**359-362.

25. Koch, W. H., and Chedekel, M. R. (1986). Photoinitiated DNA damage by melanogenic intermediates *in vitro. Photochem. and Photobiol.*, **44**:703-710.
26. D'Amato, R. J., Lipman, Z. P., and Snyder, S. H. (1986). Selectivity of the Parkinsonian Neurotoxin MPTP: Toxic metabolite MPP^+ binds to neuromelanin. *Science*, **231**:987-989.
27. Sealy, R. C. (1984). Radicals in melanin biochemistry. *Methods in Enzymol.*, **105**:479-483.
28. Arnaud, R. Perbet, G., Deflandre, and Lang, G. (1983). Electron spin resonance of melanin from hair. Effects of temperature, pH and light irradiation. *Photochem. Photobiol.*, **38**:161-168.
29. Sarna, T., and Sealy, R. C. (1984). Free radicals from eumelanins: Quantum yields and wavelength dependence. *Arch. Biochem. Biophys.*, **232**:574-578.
30. Das, K. C., Abramson, M. B., and Katzman, R. (1978). Neuronal pigments: Spectroscopic characterization of human brain melanin. *J. Neurochem.*, **30**:601-605.
31. Kollias, N., and Baqer, A. (1985). Spectroscopic characteristics of human melanin *in vivo. J. Invest. Dermatol.*, **85**:38-42.
32. Sarna, T., and Sealy, R. C. (1984). Photoinduced oxygen consumption in melanin systems. Action spectra and quantum yields for eumelanin and synthetic melanin. *Photochem. Photobiol.*, **39**:69-74.
33. Sarna, T., Pilas, B., Land, E. J., And Truscott, T. G. (1986). Interaction of radicals from water radiolysis with melanin. *Bochemica et Biophysica Acta*, **883**:162-167.
34. Prevot, H., Jr., Linden, W. A., Wehbe, S., and Lugger, A. (1970). In-vitro-Untersuchungen and Saugetierzellen uber die Strahlenschutzwirkung von Melanin. *Strahlentherapie*, **140**:753-756.
35. Johnson, B. E., Mandell, G., and Daniels, F., Jr. (1972). Melanins and cellular reactions to ultraviolet radiation. *Nature*, **235**:147-149.
36. Zhdanova, N. N., Svishchiuk, A. A., Shmyguhn, M. P., and Bondar, A. I. (1970). On the protective effect of the pigment isolated from cladosporium SP. *Microbiologia*, **36**:603-607.
37. Hill, H. Z., Huselton, C., Pilas, B., and Hill, G. J. (1987). The ability of melanins to protect against the radiolysis of thymine and thymedine. *Pigment Cell Research*, **1**:81-86.
38. Hill, H. Z., and Setlow, R. B. (1982). Comparative action spectra for pyrimidine dimer formation in Cloudman S91 mouse melanoma and EMT6 mouse mammary carcinoma cells. *Photochem. Photobiol.*, **35**:681-684.
38a. Huselton, C. A. and Hill, H. Z. (1990). Melanin photosensitizes ultraviolet light (UVC) DNA damage in pigmented cells. *Environmental and Molecular Mutagenesis* **16** in press.
39. Hill, H. Z., Peak, J. G., and Peak, M. J. (1989). Induction of DNA-protein crosslinks in melanotic Cloudman S91 mouse melanoma cells and EMT6 mouse mammary carcinoma cells by monochromatic 254 and 405 nm light. *Pigment Cell Research*, **2**: 427-430.
40. Pajak, S., Hopwood, L. E., Hyde, J. S., Felix, C. C., Sealy, R. C., Kushnaryov, V. M., and Hatchell, M. C. (1983). Melanin endocytosis by cultured mammalian cells. A model for melanin in a cellular environment. *Exp. Cell Res.*, **149**:513-526.
41. Hopwood, L. E., Swartz, H. M., and Pajak, S. (1985). Effect of melanin on radiation response of CHO cells. *Int. J. Radiat. Biol.*, **47**:531-537.
42. Kohn, K. W., Ewig, R. A. G., Erickson, L. C., and Zwelling, L. A. (1981). Measurement of strand breaks and cross-links by alkaline elution. **In**: *DNA Repair*. E. C. Friedberg, and P. C. Hanawalt, Eds. Marcel Dekker, Inc., New York, pp. 379-401.
43. Hill, H. Z., and Hill, G. J. (1987). Eumelanin causes DNA strand breaks and kills cells. *Pigment Cell Research*, **1**:163-170.

4

hnRNP PROTEIN A1 AND INSIGHT ON THE MECHANISM OF NUCLEIC ACID BINDING

Samuel H. Wilson

Laboratory of Biochemistry
National Cancer Institute
National Institutes of Health
Bethesda, MD 20892

SUMMARY

Mechanisms of protein-nucleic acid interaction are now being studied in detail for two different classes of nucleic acid-binding proteins. These classes are 1) the sequence-specific, double-stranded DNA-binding proteins, notably *Eco*RI endonuclease, phage lambda repressors Cro and CI, and cellular repressors termed trp, lac, and catabolite repressor; and 2) the sequence-nonspecific, single-stranded nucleic acid-binding proteins, one of which is the subject of this review. It is well established that sequence-specific protein-DNA recognition involves precise contacts between nucleotide residues in DNA and a "helix-turn-helix" structural motif in the protein. In the case of the sequence-nonspecific, single-stranded nucleic acid-binding proteins, extended studies of fd phage gene 5 protein and T4 phage gene 32 protein have been conducted. Both proteins have a long, lysozyme-like binding grove on the surface and rely upon tyrosine and phenylalanine residues within the grove to form hydrophobic interactions with nucleotides. As with the sequence-specific binding proteins, multiple charge-charge interactions and hydrogen bonds are also involved.

Recent studies of the mammalian single-stranded nucleic acid-binding protein, A1, have yielded new detailed information on another member of this group of sequence-nonspecific binding proteins. The binding mechanism of the A1 protein differs somewhat from the mechanism of gene 5 protein or of gene 32 protein. The A1 protein is the first mammalian single-stranded nucleic acid-binding protein to be overexpressed in *E. coli*, purified, and examined in detail. The sequence of the protein is highly conserved among vertebrates. The protein is organized into two overall domains, NH_2-terminal and COOH-terminal, that are structurally and functionally distinct. Both domains bind to nucleic acid, but the COOH-terminal domain confers positive cooperativity to A1 binding through protein-protein interaction. The COOH-terminal domain shares homology with keratin and has a novel sequence motif of repeating heptapeptide units (~7 residues), each with one tyrosine or phenylalanine residue. Interestingly, a motif of similar repeats is found in the COOH-terminal domain of eukaryotic RNA polymerase II.

INTRODUCTION

Mammalian nucleic acid helix destabilizing proteins are a class of single-stranded (ss) nucleic acid-binding proteins purified by virtue of their strong and selective binding of ssDNA-cellulose columns. The proteins were first described in 1975 (1), and from 1975 to 1983, they were typified by preparations from calf thymus (1) and mouse myeloma (2), although analogous proteins had been obtained from HeLa and a variety of other eukaryotic cell types. The calf and mouse proteins were purified as 24 and 27kDa polypeptides that are monomers at physiological salt concentrations (1,2). The purified proteins bind to ssDNA or RNA with much higher affinity than to double-stranded (ds) DNA and are capable of strongly depressing the melting temperature (Tm) of dsDNA (2-4). Binding to ssDNA is noncooperative. The binding site of each protein monomer spans seven nucleotide residues, and a length of ssDNA can be almost fully covered by the protein, since protein-protein and protein-DNA interactions interfering with binding are weak (2). Binding affinities to RNA and ssDNA are approximately equal, and the proteins can promote conformational changes in RNA secondary structure (3-5) equivalent to Tm depression of dsDNA. Hence, the proteins are properly considered "single-stranded nucleic acid-binding proteins," or nucleic acid helix destabilizing proteins (HDPs), rather than as DNA-binding proteins, exclusively (4-6).

More recent experiments exploiting immunological probing of unfractionated extracts of mammalian cells revealed that the native species of HDP has a M_r of 35,000 (7) and that HDP and the 30S heterogeneous nuclear ribonucleoprotein (hnRNP) complex protein, A1, are identical. Hence, the HDP, or A1, is involved in mRNA metabolism as a structural component of the hnRNP complex. The protein could be involved in DNA metabolism as well, since binding to ssDNA is strong and purified preparations of protein stimulate both the initial rate and processivity of DNA polymerase α, the replicative polymerase of eukaryotic cells. These effects are illustrated by experiments with purified mouse DNA polymerase α shown in Figures 1 and 2. First, the rate of α-polymerase activity on poly d(A-T) as template primer is stimulated 5- to 10-fold. Second, the processivity on poly d(T) as template is enhanced such that the early termination normally seen with this enzyme,

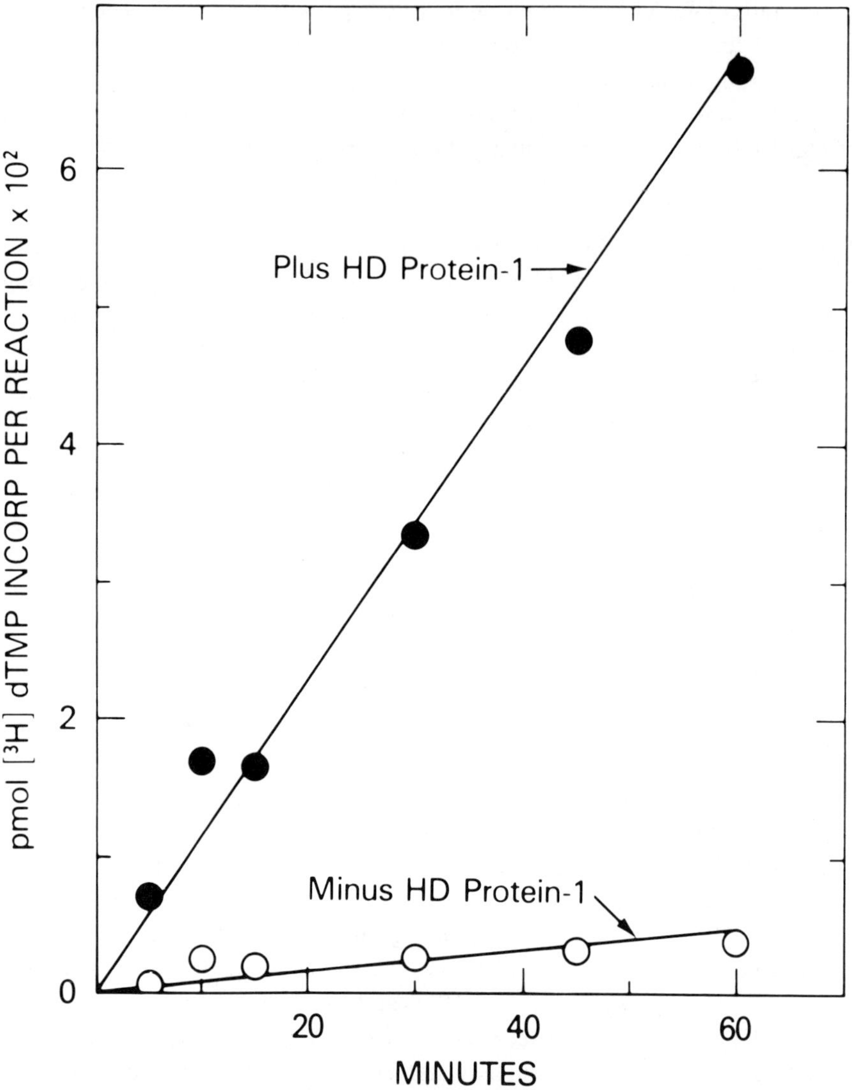

Figure 1: Effect of mouse helix-destabilizing protein-1 (HDP-1) on the rate of DNA polymerase activity of mouse α-polymerase. Incubations at 37°C contained 60 μg/ml HD protein-1, and the template-primer was poly d[A-T] (25).

after just 12 or fewer nucleotide additions to the primer, is almost completely suppressed, and the chain length of longer, processively formed products is much greater than without the binding protein. Stimulation of DNA

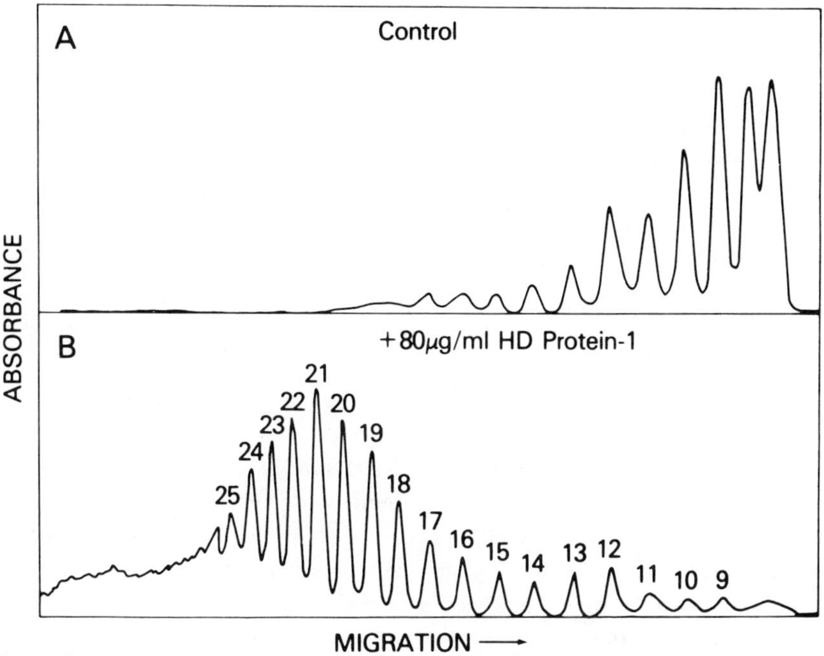

Figure 2: Effect of mouse HDP-1 on the chainlength of products formed by mouse α-polymerase. DNA polymerase products (^{32}P labeled) were separated by electrophoresis in a polyacrylamide gel under denaturing conditions and an autoradiogram was scanned by densitometry; incubations contained 80 μg/ml of template and 4 μg/ml of primer. The numbers above the peas indicate the chainlength of the 5′-OH([^{32}P]dA)$_n$ product molecules. All products were formed processively. (Reprinted from ref. 25 with permission.)

polymerase activity also has been reported by a number of other research groups (see ref. 6 for discussion). The effect is mediated through interaction with the template rather than exclusively with the enzyme, since protein amounts stoichiometric with the template and vastly exceeding the enzyme are required. There is enzyme-protein specificity in the stimulation, however, as other types of DNA polymerases are not stimulated and other binding proteins cannot substitute (6). A direct involvement of A1 in DNA metabolism, however, has not been demonstrated.

A brief introduction to the hnRNP complex is also in order. Shortly after synthesis on chromatin, hnRNA combines with a discrete set of polypeptides to form a large RNA-protein complex known as the hnRNP fibril (for discussion see refs. 11-14). Seen in an electronmicrograph, this structure resembles a fiber with beads interspersed from one end to the other. The beads correspond to RNP particles, and under appropriate conditions, these particles are released from the fibril and extracted from nuclei as individual units. The particles are characterized by sedimentation rate (~30S) and a set of 6-10 abundant polypeptides. In the HeLa cell system, for example, the seven major hnRNP core polypeptides are termed A1, A2, B1, B2, C1, C2, and C3, respectively (12,13). These so-called core polypeptides have M_r of 32,000 to 40,000, are moderately basic, and have similar amino acid compositions, including high glycine and low cysteine content and the unusual amino acid dimethylarginine. The proteins are immunologically cross-reactive and represent a family of polypeptides related by a certain amount of primary structure homology (14). These polypeptides all have blocked NH_2-termini and contain a domain of ~12,000 daltons that can be removed by mild proteolytic digestion.

The discovery of the sequence of the mRNA for the rat A1 nucleic acid-binding protein (8), along with sequencing of the analogous protein isolated from calf thymus (9,10), yielded new information about the protein and its mRNA. One key finding was confirmation that the protein deduced from the mRNA is indeed equivalent to the A1 protein of the ~30S hnRNP complex. Use of this cDNA now has facilitated studies of the protein's binding mechanism, as well as the cloning of cDNAs for related proteins, such as the A2 protein of the 30S hnRNP complex and A1 homologues of other species.

RESULTS

Cloning of A1 cDNA

Our early studies on amino acid sequence homology of subspecies of the HDP protein revealed that HDPs isolated from rodent and calf sources were strikingly similar. Therefore, we assumed that the primary structure of the calf protein could be the basis for cDNA cloning from a rodent library. (Note that the calf HDP is termed unwinding protein 1 (UP1) for historical reasons.) Thus, oligonucleotide probes corresponding to pentapeptide sequences in the NH_2-terminal and COOH-terminal portions of calf UP1 were used (8) to obtain a long cDNA from a newborn rat brain cDNA library in λgt11. The nucleotide sequence of this cDNA is shown in Figure 3. The reading frame and translation start codon of the corresponding mRNA were assigned by comparison with the amino acid sequence of calf UP1 (9).

```
                10           20           30           40           50           60
  cgctgaacgctctcatcatcctaccgtc ATG TCT AAG TCA GAG TCC CCC AAG GAA CCG GAA CAG
                               Met Ser Lys Ser Glu Ser Pro Lys Glu Pro Glu Gln  12
     70          80          90          100         110         120
CTG CGG AAG CTC TTC ATT GGA GGG CTG AGC TTC GAA ACA ACC GAC GAG AGT CTG AGG AGC
Leu Arg Lys Leu Phe Ile Gly Gly Leu Ser Phe Glu Thr Thr Asp Glu Ser Leu Arg Ser  32
     130         140         150         160         170         180
CAT TTT GAG CAA TGG GGA ACA CTC ACC GAC TGT GTG GTA ATG AGA GAT CCA AAC ACC AAA
His Phe Glu Gln Trp Gly Thr Leu Thr Asp Cys Val Val Met Arg Asp Pro Asn Thr Lys  52
     190         200         210         220         230         240
AGA TCC AGA GGC TTT GGG TTT GTC ACA TAT GCC ACT GTG GAG GAA GTG GAT GCT GCC ATG
Arg Ser Arg Gly Phe Gly Phe Val Thr Tyr Ala Thr Val Glu Glu Val Asp Ala Ala Met  72
     250         260         270         280         290         300
AAT GCA AGA CCA CAC AAA GTG GAT GGA AGA GTT GTG GAA CCT AAG AGA GCT GTG TCA AGA
Asn Ala Arg Pro His Lys Val Asp Gly Arg Val Val Glu Pro Lys Arg Ala Val Ser Arg  92
     310         320         330         340         350         360
GAA GAT TCT CAG AGA CCA GGT GCC CAC TTA ACT GTG AAG AAG ATC TTT GTT GGC GGT ATT
Glu Asp Ser Gln Arg Pro Gly Ala His Leu Thr Val Lys Lys Ile Phe Val Gly Gly Ile 112
     370         380         390         400         410         420
AAA GAA GAC ACT GAA GAA CAT CAC CTA CGA GAT TAT TTT GAG CAG TAT GGG AAA ATT GAA
lys Glu Asp Thr Glu Glu His His Leu Arg Asp Tyr Phe Glu Gln Tyr Gly Lys Ile Glu 132
     430         440         450         460         470         480
GTG ATT GAA ATT ATG ACT GAC AGA GGC AGT GGA AAA AAG AGG GGA TTT GCG TTT GTC ACC
Val Ile Glu Ile Met Thr Asp Arg Gly Ser Gly Lys Lys Arg Gly Phe Ala Phe Val Thr 152
     490         500         510         520         530         540
TTT GAT GAC CAT GAC TCT GTG GAT AAG ATT GTT ATT CAG AAA TAC CAT ACT GTG AAT GGC
Phe Asp Asp His Asp Ser Val Asp Lys Ile Val Ile Gln Lys Tyr His Thr Val Asn Gly 172
     550         560         570         580         590         600
CAC AAC TGT GAA GTA AGA AAG GCT CTG TCG AAA CAA GAG ATG GCT AGT GCT TCA TCC AGC
His Asn Cys Glu Val Arg Lys Ala Leu Ser Lys Gln Glu Met Ala Ser Ala Ser Ser Ser 192
     610         620         630         640         650         660
CAG AGA GGT CGA AGT GGT TCC GGA AAC TTT GGT GGT GGT CGT GGA GGT GGT TTC GGT GGC
Gln Arg Gly Arg*Ser Gly Ser Gly Asn Phe Gly Gly Gly Arg Gly Gly Gly Phe Gly Gly 212
     670         680         690         700         710         720
AAT GAC AAT TTT GGT CGA GGA GGG AAC TTC AGT GGT CGT GGT GGC TTT GGT GGC AGC CGT
Asn Asp Asn Phe Gly Arg Gly Gly Asn Phe Ser Gly Arg Gly Gly Phe Gly Gly Ser Arg 232
     730         740         750         760         770         780
GGT GGT GGT GGA TAT GGT GGC AGT GGG GAT GGC TAT AAT GGA TTT GGC AAT GAT GGA AGC
Gly Gly Gly Gly Tyr Gly Gly Ser Gly Asp Gly Tyr Asn Gly Phe Gly Asn Asp Gly Ser 252
     790         800         810         820         830         840
AAT TTT GGA GGT GGT GGA AGC TAC AAT GAT TTT GGC AAT TAC AAC AAC CAG TCA TCA AAT
Asn Phe Gly Gly Gly Gly Ser Tyr Asn Asp Phe Gly Asn Tyr Asn Asn Gln Ser Ser Asn 272
     850         860         870         880         890         900
TTT GGA CCG ATG AAA GGA GGA AAC TTT GGA GGC AGG AGC TCT GGC CCT TAT GGT GGT GGA
Phe Gly Pro Met Lys Gly Gly Asn Phe Gly Gly Arg Ser Ser Gly Pro Tyr Gly Gly Gly 292
     910         920         930         940         950         960
GGC CAG TAC TTT GCT AAA CCA CGA AAC CAA GGT GGC TAT GGA GGT TCC AGC AGC AGC AGT
Gly Gln Tyr Phe Ala Lys Pro Arg Asn Gln Gly Gly Tyr Gly Gly Ser Ser Ser Ser Ser 312
     970         980         990        1000       1010       1020       1030
AGC TAT GGC AGT GGC AGG AGG TTC TAA ttacagccaggaaacaaagcttagcaggagagccagagaagtg
Ser Tyr Gly Ser Gly Arg Arg Phe ***
    1040      1050      1060      1070      1080      1090      1100      1110
acagggaagctacaggttacaacagatttgtgaactcagccaagcacagtggtggcagggcctagctgctacaaagaag
     1120      1130      1140      1150      1160      1170      1180      119
acatgttttagacaatactcatgtgtgtgggcaaaaactccaggactgtatctgtgactaattgtataacaggttattt
0      1200      1210      1220      1230      1240      1250      1260      12
tagtttctgttctgtggaaagtgtaaagcattccaacgaagggttttactgtagaccttttcacccatgctgttgatt
70     1280      1290      1300      1310      1320      1330      1340      1
gctaaatgtaaaagtctgatcatgacgctgaataaatgtgtcttttttttttttttaatgtgctgtgtaaagttagt
350      1360      1370      1380      1390      1400      1410      1420
ctactctgaagccatcttggtaaacttccccaacagtgtgaagttagaattccttcagggtggtgccaaattccatttg
1430     1440      1450      1460      1470      1480      1490      1500
gaatttatttatggttgcttgggtggagaagccattgtcttcaaaaaccttggtgttgctaaactgccagttactgttg
 1510      1520      1530      1540      1550      1560      1570      1580
taacttaatgagtttcaccatttaaagggtcatccaagcaagatcacaatttggttataaaatggttgttggacaccta
  1590      1600      1610      1620      1630      1640      1650      1660
tgaagagaaaattgaataacggtctcagataaaataagagatgggaatgaagcttgtgtatcatccattatcatgtgta
   1670      1680      1690      1700
ctcaataaacgatttaattctcttgaatggaaaaaaaaaa
```

Figure 3: The nucleotide sequence of the A1 (λHDP-182) cDNA insert and the deduced sequence of the 320-amino acid, 34,218-dalton protein corresponding to the single long open reading frame. The coding strand is shown. The first amino acid residue in the deduced protein corresponds to the first AUG from the 5′ end; the last amino acid residue is adjacent to the first in-phase termination codon. Residues 2 through 196 of the deduced protein are identical to the reported sequence of calf thymus UP1, the end of which is marked (*). In the 3′ untranslated region, a putative mRNA processing signal (AATAAA) and repeated sequences around a T sequence are noted. In the 5′ untranslated region, a sequence complementary to a sequence in 18S rRNA is noted. (Reprinted from ref. 15 with permission.)

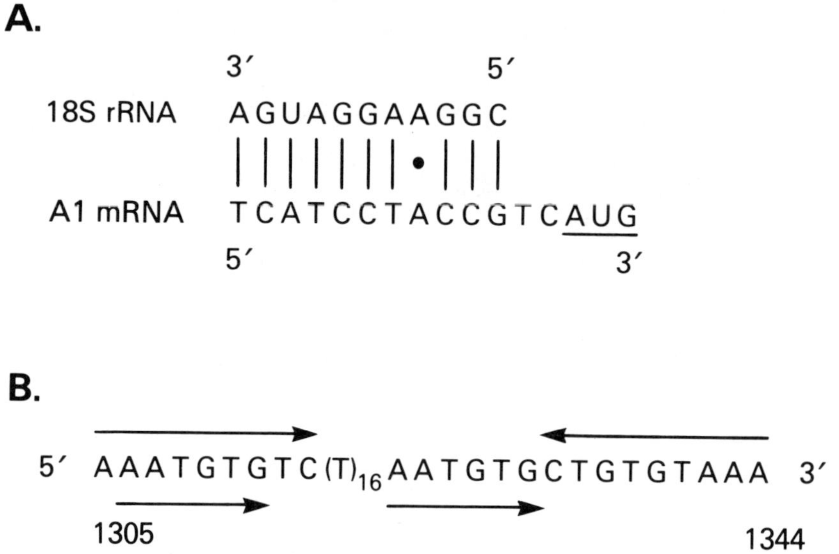

Figure 4: Diagrams illustrating sequences in the 3′ and 5′ untranslated regions of the A1 mRNA. (A) Sequence homology between the 3′ end of the 18S rRNA and a sequence in the 5′ untranslated region of the HDP cDNA; note that the A1 start codon is listed as AUG. (B) Summary of direct and inverted repeats surrounding the poly T sequence in the A1 cDNA. Numbering corresponds to the cDNA sequence in Figure 3. (Reprinted from ref. 15 with permission.)

The A1 mRNA

The start codon of the cDNA is the first AUG from the 5′ end corresponding to residues 29, 30, and 31. The triplet adjacent to the AUG encodes a serine residue, and a blocked NH_2-terminal serine residue is found in the protein purified from tissue (9). The distance from the AUG start codon to the first in-phase termination codon is 988 bases. The sequence from this termination codon to the start of the poly(A) tail at the 3′ end contains a 16-residue $d(T)_n$ stretch flanked by direct repeats of the hexanucleotide sequence AATGGT and by the palindromic sequence AAATGTGTC (Figure 4). Similar 3′ untranslated region sequences are seen also in histone and high mobility group protein mRNAs. Since the $d(T)_n$ sequence is flanked by direct repeats, the overall sequence of homopolymer and flanking nucleotides is similar to a transposable element.

The HDP mRNA has interesting structural features in the 5′ untranslated region as well. This region contains an 11-residue sequence complementary to a sequence in the 3′ end of 18S rRNA (Figure 4). This sequence, which

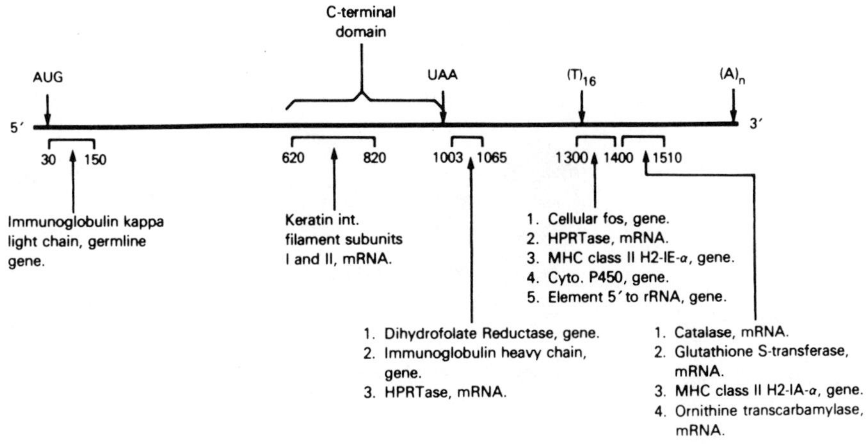

Figure 5: Scheme summarizing partial nucleotide sequence homologies between the A1 cDNA and sequences in the GeneBank Data Bank. Numbering corresponds to the cDNA sequence in Figure 3. All homologies illustrated were significant at the level of 5 or more standard deviation units above mean. The SRCHN program of D.J. Lipman and W.J. Wylbur, NIH, was used with K-tuple size = 4, window size = 20, and gap penalty = 4.

corresponds to residues 18-26 of the cDNA, may be involved in control of translation.

A computer search for rodent sequences homologous to the A1 mRNA revealed a number of partial, albeit significant, homologies. None of the homologous sequences identified formed a perfect match with the A1 cDNA, but all were significant at the level of five or more standard deviation units above the mean. With the exception of two cases, all of the homologies corresponded to the 3′ untranslated region of the A1 mRNA (Figure 5). The 100-residue region including and surrounding the 16 d(T) sequence shares homology with the cellular *fos* gene, with a mobile element 5′ of rRNA genes, and with three other genes of no obvious interrelationship. Similarly, two other relatively discrete segments corresponding to the 3′ untranslated region share homology with four other genes; none of these genes or protein products have an obvious interrelationship.

Figure 5 also illustrates that a 200-nucleotide segment of the coding region of the A1 mRNA shares partial homology with the mRNA for keratin intermediate filament subunits I and II. This segment of the cDNA falls near

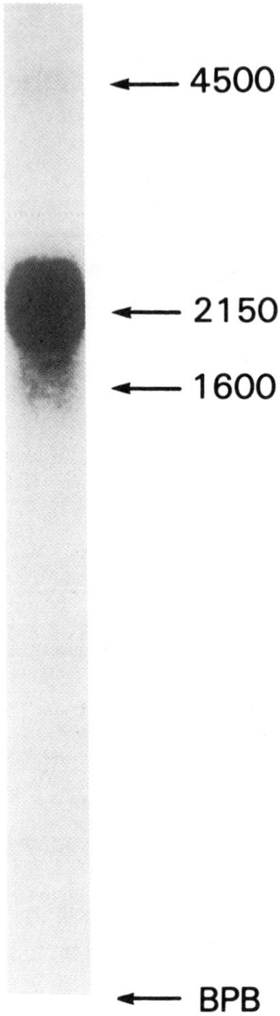

Figure 6: Autoradiogram showing results of Northern blot analysis of newborn rat brain poly(A)$^+$ RNA. RNA was prepared by the guanidine-HCl method, and a sample containing 4 μg was electrophoresed in a 1.5% agarose gel containing 50% formamide. The 1397-bp RI fragment was ^{32}P-labeled by nick translation and used as probe. Exposure was for 1 h. Base sizes indicated were relative to markers of rat brain rRNA, 18S (1950) and 28S (5000). *BPB*, bromphenol blue. (Reprinted from ref. 8 with permission.)

the end of the open reading frame and corresponds to the COOH-terminal portion of the deduced protein. As will be described below, the homology with keratin mRNA extends to the protein level also and is extensive.

Northern blot analysis of newborn rat brain poly(A)$^+$ RNA was conducted using as probe a fragment of the cDNA containing the first 1397 base pairs. After only a 1-hour exposure of the autoradiogram, a strong predominant signal was observed corresponding to approximately 2100 bases (Figure 6). A higher M_r species of ~4500 bases and a smaller species of ~1600 bases were noted also. Similar results were obtained with total RNA, instead of poly(A)$^+$ RNA, and when the denaturation of RNA was performed in the presence of glyoxal, instead of formaldehyde and formamide (data not shown). The amount of the 2.1-kb transcript is approximately 0.5% of total poly(A)$^+$ mRNA.

The abundance of the 2.1-kb transcript was examined as a function of the cell growth curve using contact-inhibited 3T3 mouse fibroblast and Chinese hamster ovary cells. The mRNA did not fluctuate in abundance or size throughout the growth curve; the transcript, therefore, appears typical of a cellular housekeeping gene. Furthermore, the abundance of the A1 mRNA is not altered by stress such as heat shock, treatment with a variety of DNA damaging agents, or exposure to ultraviolet or ionizing radiation.

The A1 Protein: Theoretical Considerations

Sequencing of the cDNA yielded clues about the structure of the A1 protein not available from earlier work with the purified protein itself. The open reading frame of the cDNA, for example, indicated that the mRNA encodes a 34,215-dalton polypeptide of 320 amino acids. Computer-derived secondary structure plots of the 34,215-dalton protein clearly indicate two major structural domains and a sharp transition between these domains occurring between amino acid residues 190 and 200 (Figure 7). The NH_2-terminal domain of the 34,215-dalton protein appears to be tightly structured with prediction of four α-helix regions, seven β-sheet regions, and alternating hydrophobic and hydrophilic regions. The primary amino acid sequence of this domain is identical between the rat cDNA-derived protein and the proteins purified from calf thymus and HeLa cells, indicating strong selective pressure for conservation. The COOH-terminal domain of the protein has about 125 amino acids and is structurally flexible due to the high content of amino acid residues with β-turn and random coil prediction. This domain does not have regions of α-helix prediction and is neither hydrophilic nor hydrophobic (Figure 7). The high content of serine residues provides ample sites for phosphorylation; it has been reported that A1 isolated from hnRNP particles is phosphorylated (12).

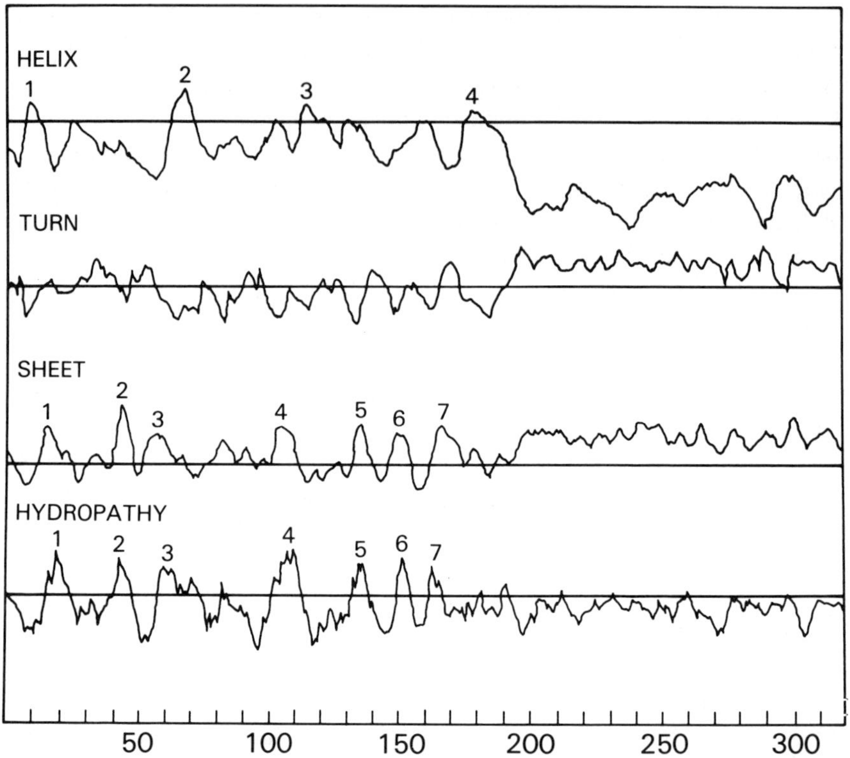

Figure 7: Computer-derived secondary structure and hydropathy plots for newborn rat brain A1. Amino acid residue numbers are displayed relative to colinear secondary predictions and a hydropathy plot. For predictions of conformation, all values plotted *above* the line indicate a prediction for that conformational feature. All values *above* the line for the hydropathy plot indicate hydrophobic characters, whereas values *below* the line indicate hydrophilic character of the polypeptide.

Analysis of the primary structure of the COOH-terminal domain of the deduced protein revealed that it contains 16 repeating oligopeptide units (Table 1), beginning with amino acid residue 202. These units consist of phenylalanine or tyrosine residues separated by five to eight amino acid residues of likely random coil structure; seven of these repeating units contain an arginine residue.

The pattern of repeating units in the peptide sequence suggests an evolutionary history of gene duplications with subsequent deletions and changes, as illustrated by the alignment of the right-hand column of Table 1. The suggestion of gene duplication is also evident from inspection of the cor-

Table 1:

Two primary structure alignments of oligopeptides in the COOH-terminal domain of the 34,215-dalton HDP based upon spacing of phenylalanine or tyrosine residues

Alignment[A] Illustrating Repeating Units	Alignment[A] Illustrating Conservation of Sequences
↓	↓ ↓ ↓
FGGGRGGG	GNFGGGRG
FGGNDN	GGFGG
FGRGGN	DNF––GRG
FSGRGG	GNF––SRG
FGGSRGGGG	GGFGGSRGG
YGGSGDG	GGYGGS–GDGYNGFGN
FGNDGSN	SNFGG––GGSYNDFGN
FGGGGS	SNFGPNKG
YNDFGN	GNFGG–RSS
YNNQSSN	GPYGGG––GQY––F
FGPMKGGN	GGYGGS–SSS
FGGRSSGP	SSYGSGRR––F
YGGGGQY	
FAKPRNQGG	
YGGSSSSSS	
YGSGRRF	
Consensus Sequence:	
FGGGGS	GNFGG(S/G)RG

[A]Alignment of phenylalanine (F) or tyrosine (Y) residues is illustrated by arrows. The oligopeptides are listed in the order of their occurrence in the COOH-terminal domain starting with F 202 and extending to the COOH terminus F.

responding nucleotide sequence. The oligonucleotide CAATTTTGG is present twice, and the related sequences AAACTTTGG and ATTTTGG are present two times each as direct repeats. The sequence ATTTTGG and a second closely related sequence are present once each as inverted repeats.

Table 2: Summary of COOH-terminal domain oligopeptide repeats in nucleic acid proteins

Protein	Number and Length of Repeat Unit		Consensus Sequence and Number	Reference
	Number	Length		
Mouse RNA Pol II largest subunit	52	7	TyrSerProThrSerProSer 22	23
Yeast RNA Pol II largest subunit	26	7	TyrSerProThrSerProSer 17	24
Al	16	6 to 9	PheGlyGlyGlyGlySer 1	8

The functional significance of the structural domains in the A1 remains to be determined. It is interesting to note, however, that in both yeast and mouse RNA polymerase II the largest subunit contains a COOH-terminal domain with repeating heptapeptide units, each with one aromatic residue (Table 2). The primary structure of the consensus sequence of the RNA polymerase II subunit repeats is much more represented among the various repeats than in the case of the consensus sequence for the A1 repeats. The periodicity of the aromatic residues, however, is a common feature of the repeating structures of both proteins. This and other structural features of the repeats provide interesting possibilities for binding to nucleic acids. A plausible conformation for the RNA polymerase II repeating units and the close interaction of this conformation with a single-stranded nucleic acid are illustrated with the computer-derived space-filling model shown in Figure 8. This model (15) shows binding of a theoretical polymer with 17 of the consensus repeat units of RNA polymerase II (Table 2) to doubled-stranded B-DNA. Each seven-residue unit has two highly predicted β-turns leading to a helical configuration. The aromatic tyrosine residue in each unit is on the inside of the helix and would lie very close to the nucleic acid bases. Although it is not visible in this illustration, the alignment of the aromatic residue and each base pair is perpendicular, not planar. Thus, this structure proposed with 1:1 interaction between nucleic acid base residues and aromatic residues would explain the periodicity of aromatic residues in the protein. Model building showed that this close fit between protein and nucleic acid would be disrupted by phosphorylation of serines, threonines, or tyrosines due to steric hindrance and electrostatic repulsion. The protein helix shown in Figure 8 has the same pitch and number of repeats per turn as B-DNA or RNA-DNA hybrids and some types of single-stranded nucleic acids.

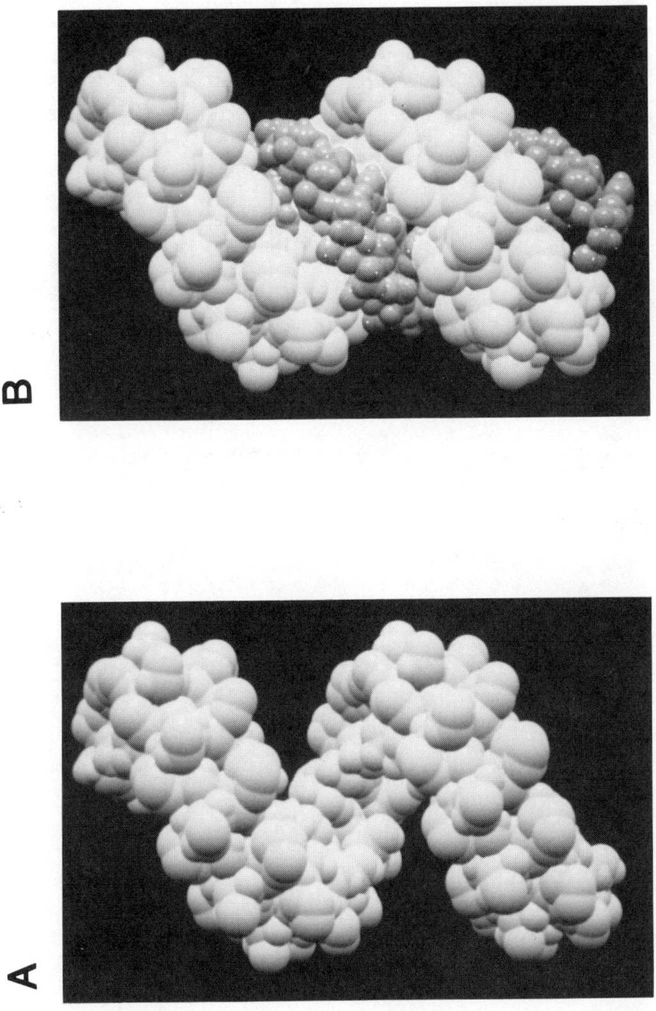

Figure 8: Computer-derived model of hypothetical structure of RNA polymerase II COOH-terminal domain and its interaction with a nucleic acid. A protein (darker shaded areas at middle and bottom of photograph B) composed of 17 repeats of the consensus sequence (Ser-Pro-Thr-Ser-Pro-Ser-Tyr) assumes a conformation that forms a helix with 11 repeats per turn and 2.6 translations per repeat, which is the same as B-DNA or DNA-DNA hybrids. Tyr groups are oriented toward the center of the helix. (A) DNA alone. (B) A strand of B-DNA is shown interacting with the RNA polymerase domain. The phosphate-sugar backbone is visible. The bases are covered by the protein and interact with the Tyr groups. It is possible that the protein conformation could be altered to interact with other nucleic acid structures such as that found for single-stranded polycytidylic acid or double-stranded A-DNA or DNA-RNA hybrids. (Reprinted from ref. 15 with permission.)

```
                              10                  20                  30
A1(1)        4    E S P K E P E Q L R K L F I G G L S F E T T D E S L R S H * * *
A1(2)       95    Q R P G A H L T V K K I F V G G I K E D T E E H H L R D Y * * *
UP2        163    K A M K T K E P V K K I F V G G L S P D T P E E K I R E Y * * *
C1           7    N K T D P R S M N S R V F I G N L N T L V V K K S N D E A * * *
NUCL(1)    298    V E G S E S T T P F N L F I G N L N P N K S V A E L K V A * * *
NUCL(2)    384    R D S K K V R A A R T L L A K N L S F N I T E D E L K E V * * *
NUCL(3)    476    K N S T W S G E S K T L V L S N L S Y S A T E E T L Q E V * * *
NUCL(4)    562    S P N A R S Q P S K T L F V K G L S E D T T E E T L K E S * * *
FL70       270    D P N A Q G D A F K T L F V A R V N Y D T T E S K L R R E * * *
U2B"(1)      1          M D I R P N H T I Y I N N M N D K I K K E E L K R S L Y A
U2B"(2)    142    P Q V P D Y P P N Y I L F L N N L P E E T N E M M L S * * * * M
DMp9(1)     21    D S I T E P E H M R K L F I G G L D Y R T T D E N L K A H * * *
DMp9(2)    113    D S P N A G A T V K K L F V G A L K D D H D E Q S I R D Y * * *
YPAB(1)     28    E S Q S V E N S S A S L Y V G D L E P S V S E A H L Y D I * * *
YPAB(2)    117    D P S L R K K G S G N I F I K N L H P D I D N K A L Y D T * * *
YPAB(3)    209    Q L E E T K A H Y T N L Y V K N I N S E T T D E Q F Q E L * * *
YPAB(4)    312    L E K M A K Y Q G V N L F V K N L D D S V D D E K L E E E * * *

                                40                  50                  60
A1(1)       33    * F E Q W G T L T D C V V M R D P N T K R S R G F G F V T Y A T
A1(2)      124    * F E Q Y G K I E V I E I M T D R G S G K K R G F A F V T F D D
UP2        192    * F G G F G E V E S I E L P M D N K T N K R R G F C F I T F N Q
C1          36    I F S K Y G K I V G C S V * * * * * * * * H K G F A F V Q Y V N
NUCL(1)    327    I S E P F A K N D L A V V * * D V R T G T N R K F G Y V D F E S
NUCL(2)    413    * F E D A L E I R L V * * * * * S Q D G K S K G I A Y I E F K S
NUCL(3)    505    * F E K A T F I K V P Q * * * * N Q Q G K S K G Y A F I E F A S
NUCL(4)    591    * F E G S V R A R I V T * * * D R E T G S S K G F G F V D F N S
FL70       309    * F E V Y G P I K R I H M V Y S K R S G K P R G Y A F I E Y E H
U2B"(1)     30    L F S Q F G H V V D I V A L * * * K T M K M R G Q A F V I F K E
U2B"(2)    170    L F N Q F P G F K E V R L V * * * * * P G R H D I A F V E F E N
DMp9(1)     51    * F E K W G N I V D V V V M K D P R T K R S R G F G F I T Y S H
DMp9(2)    143    * F Q H F G N I V D I N I V I D K E T G K K R G F A F V E F D D
YPAB(1)     48    * F S P I G S V S S I R V C R D A I T K T S L G Y A Y V N F N D
YPAB(2)    147    * F S V F G D I L S S K I A T D * E N G K S K G F G F V H F E E
YPAB(3)    239    * F A K F G P I V S A S L E K D A D * G K L K G F G F V N Y E K
YPAB(4)    342    * F A P Y G T I T S A K V M R T * E N G K S K G F G F V C F S T
YSSB-1                                                            A F V T F X I

                              70                  80                  90
A1(1)       64    V E E V D A A M N A R P H K * V D G R V * V E P K R A V S R E D
A1(2)      155    H D S V D K I V I Q K Y H T * V N G H N * C E V R K A L S K Q E
UP2        223    E E P
C1          60    E R N A R A A V A G E N G R M I A G Q V * L D I N L A A E P K V
NUCL(1)    357    A E D L E K A L E L T G L K * V F G N E * I K L E K P K P K G
NUCL(2)    439    E A D A E K N L E E K Q G A E I D G R S * V S L Y Y T G E K G Q
NUCL(3)    532    F E D A K E A L N S C N K M E I E G R T * I R L E L Q G P R G
NUCL(4)    619    E E D A K A A K E A M E D G E I D G N K * V T L D W A K P K G E
FL70       340    E R D M H S A Y K H A D G K K I D G R R * V L V D V E P G R T V
U2B"(1)     59    L G S S T N A L R Q L Q G F P F Y G K P * M R I Q Y A K T D S D
U2B"(2)    197    D G Q A G A A R D A L Q G F K I T P S H A M K I T Y A K K
DMp9(1)     81    S S M I D E A Q K S R P H K * I D G R V * V E P K R A V P R Q D
DMp9(2)    173    Y D P V D K V V L Q K Q H Q * L N G K M * V D V K K A L P K Q N
YPAB(1)     78    H E A G R K A I E Q L N Y T P I K G R L * C R I M W S Q R
YPAB(2)    176    E G A A K E A I D A L N G M L L N G Q E * I Y V A P H L S R K E
YPAB(3)    268    H E D A V K A V E A L N D S E L N G E K * L Y V G R A Q K K N E
YPAB(4)    371    P E E A T K A I T E K N Q Q I V A G K P * L Y V A I A Q R K D V
YSSB-1            E N V
```

Figure 9: Alignment of eukaryotic RNA-binding protein sequences to show protein-protein as well as internal repeat sequence homologies. Residue numbers on top are for the A1 protein sequence. Proteins and sequence references are as follows: *A1*, hnRNP from rat or calf thymus (8,11); *C1*, hnRNP from HeLa (17); *UP2*, partial cDNA sequence for a protein originally thought to be a type C hnRNP (26) that has sequence homology to the 40kDa calf thymus UP2 protein (10); *NUCL*, nucleolin, the nucleolar C23 protein from rat and hamster (27); *FL70*, the U1 snRNP-associated 70kDa protein from HeLa (28); *U2B"*, the U2 snRNA-associated B" protein from HeLa (29); *DMp9, Drosophila* clone p9 cDNA sequence (30); *YPAB*, yeast poly(A)-binding protein (18,31,32). The 4 phenylalanine residues in A1 found to be cross-linked to [^{32}P]p(dT)s are *circled* (see below). Reprinted with permission from ref. 116.

The periodicity of aromatic residues and predictions of β-turns in the COOH-terminal domain of A1 indicate that A1 also could bind to nucleic acids in a manner similar to that depicted in Figure 8. Unfortunately, the A1 repeating units are difficult to model because they differ among themselves more than the repeats in RNA polymerase II (Table 2) and they contain many glycine residues, which should give the structure a great deal of conformational freedom. The possibility that these unusual COOH-terminal domain structures bind specifically to certain types of DNA or RNA structures or other cellular proteins can be evaluated experimentally after appropriate isolation of large amounts of the relevant proteins.

Examination of the amino acid sequence of the NH_2-terminal domain reveals two adjacent, 92-residue stretches of internal sequence homology such that when residues 3-93 are aligned with residues 94-184, 32% of the amino acids are identical and an additional 39% of the existing differences could result from single base substitutions. As will be described below, we have proposed that each of these internal repeats represents an independently folded, globular domain that functions in nucleic acid binding (16).

There is a degree of sequence similarity between segments of A1 and many other eukaryotic RNA-binding proteins. Thus, there are homologies between A1 and RNA proteins binding as noted previously by Swanson *et al.* (17) and Adam *et al.* (18): yeast poly(A)-binding protein, the Cl and C2 hnRNP, the UP2 hnRNP, and the nucleolar RNA-binding protein, nucleolin. There are four regions of similarity to A1 in the amino acid sequence of nucleolin, two regions in the U2 snRNP-associated B″ protein, one region in the Ul snRNP-associated 70,000-dalton protein, and two regions in the *Drosophila* p9 protein. There is also a protein in the yeast, SSB-l, with similarity to A1, suggesting that perhaps this protein has RNA-binding function. The alignment is shown in Figure 9. From this alignment, it is immediately apparent that the "RNP consensus sequence" of Adam *et al.* (18) is highly conserved and is not the only portion of the approximately 90-amino acid RNA-binding domain that is common to most of the proteins. Positions 14-20, 29, 33, 36, 37, 70, 78, 80, and 91 of A1 are also examples of well-conserved residues. Positions 14-20 are referred to as RNP consensus sequence 2, whereas the sequence at positions 54-60 is RNP consensus sequence 1.

Preparation of Reagent Quantities of A1 Protein

We undertook expression of the rat A1 cDNA in *E. coli* in order to obtain sufficient amounts of the unfused 34kDa protein for detailed studies of structure-function relationships (19). The strategy of obtaining large amounts of a protein by cDNA-based production in *E. coli* is now well established. Yet, in

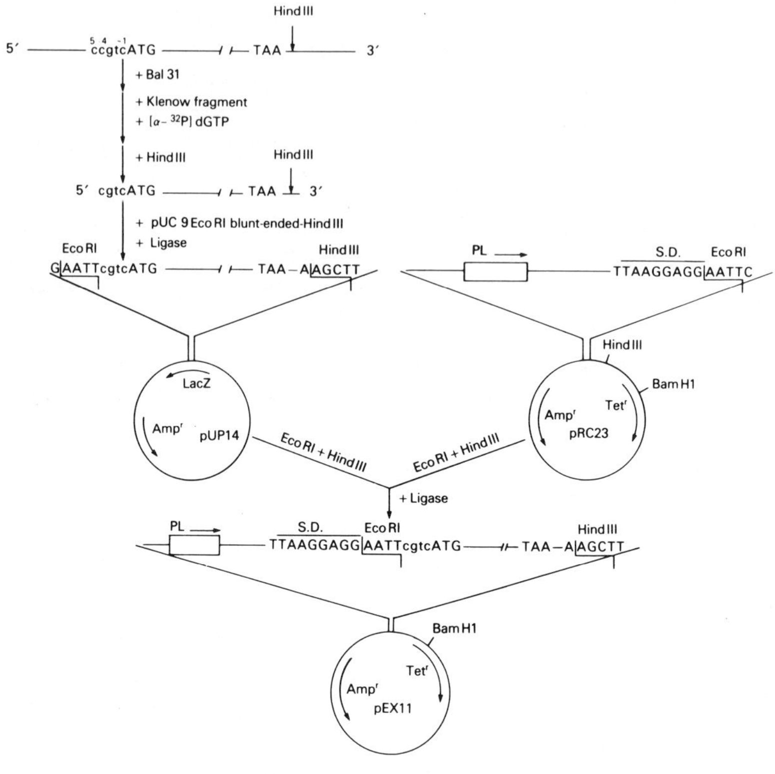

Figure 10: Construction of expression plasmid pEX11 for overproduction of A1 (19). An *Eco*RI fragment containing the rat A1 cDNA insert was resected with Ba131. Products were blunt-ended with Klenow fragment and only one dNTP (dGTP) and then inserted in pUC9. One of the isolates had the desired sequence noted and was named pUP14. The *Eco*RI-*Hind*III fragment from pUP14 was subcloned in pRC23, yielding final the plasmid, pEX11. Pl, λ left promoter; S.D., Shine-Dalgarno site; Amp, ampicillin; Tet, tetracycline.

most cases, it remains a formidable experimental achievement to obtain appropriate modification and subcloning of a coding sequence, overexpression, and purification of large amounts of undegraded protein. In the case of the A1 protein, with the full-length A1 cDNA, we obtained desirable subcloning and overexpression in a straightforward fashion. However, the complete purification of the recombinant A1 protein in undegraded form proved difficult.

More specifically, the protein was overexpressed from the open reading frame of the cDNA using the λPL promoter-based bacterial expression system pRC23 (20) (Figure 10). Our placement of the cDNA in the vector positioned the ribosome-binding site of the vector eight bases upstream of the initiation

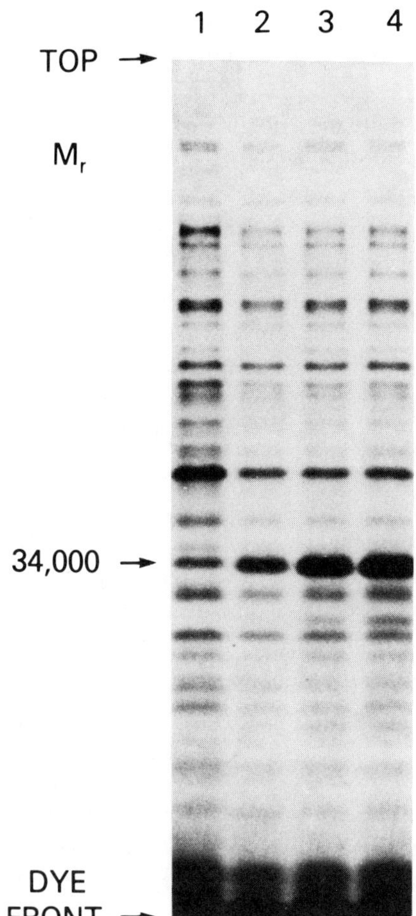

Figure 11: Overexpression of A1 in *E. coli* RR1 carrying pEX11 (19). Transformed cells were grown at 30°C to A560=0.2. The temperature was raised to 42°C. Bacterial lysates were prepared and analyzed by SDS-polyacrylamide gel (12% acrylamide, 0.325% bisacrylamide) electrophoresis. A photograph of the Coomassie blue-stained gel is shown. Lane 1, uninduced and incubated for 3 h at 30°C ; lanes 2, 3, and 4, induced and incubated for 1, 2, and 3 h, respectively, at 42°C.

codon in the cDNA (19). Cells transformed with the expression plasmid had large quantities of a new 34kDa protein (Figure 11), and immunoblotting experiments with anti-A1 antibody revealed that this protein contained a reactive epitope. Using a simple purification procedure designed to obviate proteolysis, we could obtain 50 mg of 34kDa A1 protein from the soluble extract of 10 g of pelleted cells in 10-12 hours. The final preparation of A1 is homogeneous by amino acid sequencing and by SDS-polyacrylamide gel electrophoresis, and as expected, the purified protein itself is immunoreactive with anti-A1 antibody.

The amino acid composition of purified recombinant A1 concurs with that predicted from the nucleotide sequence of the cDNA, and sequencing of the protein demonstrated that residues 1-25 are in exact agreement with residues 2-26 of the sequence predicted from the cDNA. The purified protein does not contain the cDNA-predicted NH_2-terminal methionine. Enzymatic cleavage of the protein was conducted with endoproteinase *Lys-c*. The sequence at the COOH-terminal end of the protein was determined by sequencing the only endoproteinase *Lys-c* peptide that did not contain lysine; it matches the COOH-terminal sequence predicted from the cDNA nucleotide sequence. Several additional endoproteinase *Lys-c* peptides also were sequenced. These peptides, together with the NH_2- and COOH-terminal sequences described above, account for ~40% of the complete A1 sequence. Finally, the amino acid compositions of the other *Lys-c* peptides were determined and were found to match peptides predicted from the cDNA sequence. Taken together, the results indicated a perfect match between the recombinant A1 protein sequence and the sequences predicted from the cDNA at 90% of the residues (Table 3). Note that numbering begins with Ser, not with Met, from the cDNA sequence (19).

Domain Structure and ssDNA-cellulose Binding

Secondary structure predictions suggest that A1 is a "two-domain" protein, exhibiting a sharp transition at residues 195-200 between the globular NH_2-terminal domain and the randomly structured COOH-terminal domain. Three preparations of the NH_2-terminal domain peptide were obtained for comparison with intact A1. These preparations were 1) the NH_2-terminal domain isolated from calf thymus, termed UP1; 2) the NH_2-terminal domain obtained by *in vitro* trypsinization of recombinant A1, termed P24*; and 3) a recombinant UP1 obtained by subcloning a truncated A1 cDNA into an appropriate expression vector.

The initial structure-function experiments with recombinant A1 involved the examination of the ssDNA-cellulose column chromatographic behavior of the intact protein and of products of limited proteolysis. Light trypsinization was found to convert the 34kDa A1 protein to species of 32, 30, 27, and 24 kDa. When a mixture of the 34kDa protein and all four of these truncated species was subjected to NH_2-terminal sequence analysis, only one NH_2-terminal sequence was found corresponding to that of the intact species. This indicated that conversion to the smaller species was due to truncation at the

Table 3:

Match between cDNA-predicted sequence and the determined sequence of recombinant A1

SKSESPKGPEQLRKLFIGGLSFETTDESLRSHFEQWGTLTDCVVMRDPN

TKRSRGFGFVTYATVEEVDAAMNARPHKVDGRVVEPKRAVSREDSQRPGA

HLTVKKIFVGGIKEDTEEHHLRDYFEQYGKIEVIEIMTDRGSGKKRGFAF

VTFDDHDSVDKIVIQKYHTVNGHNCEVRKALSKQEMASASSSQRGRSGSG

NFGGGRGGGFGGNDNFGRGGNFSGRGGFGGSRGGGGYGGSGDGYNGFGND

GSNFGGGGSYNDFGNYNNQSSNFGPMKGGNFGGRSSGPYGGGGQYFAKPR

NQGGYGGSSSSSSYGSGRRF

The cDNA-predicted sequence is shown. The lighter line (upper) indicates residues confirmed by amino acid composition analysis of Lys-C peptides. The heavier line (lower) indicates residues confirmed by sequence analysis.

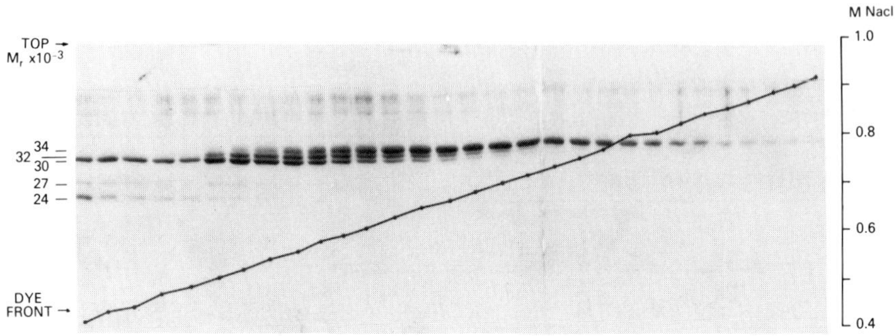

Figure 12: DNA-cellulose column chromatographic analysis of 34kDa A1/HDP and four products of limited proteolysis. SDS-polyacrylamide gel electrophoresis of column fractions was conducted, and a photograph of the Coomassie blue-stained gel is shown. A mixture of A1 and four species of proteolytic digestion was loaded on an ssDNA-cellulose column, and the column was washed with buffer containing 0.4 M NaCl. Elution was performed with a linear gradient of 0.4-1 NaCl in buffer. The two minor bands migrating in the gel between the top and 34kDa A1/HDP were not from the column samples but were contaminants occasionally seen in our gel analysis (19).

COOH-terminal end only. This mixture of A1 and truncated species was chromatographed on a ssDNA-cellulose column, and elution was performed with a linear gradient of 0.4-1 M NaCl (Figure 12). The two smaller species, 24 and 27 kDa, emerged from the column just after the start of the gradient. The 30, 32, and 34kDa species emerged successively as the NaCl concentration in the gradient increased. In additional experiments, the 24kDa species was produced by trypsin digestion of recombinant A1 as described by Kumar *et al.* (21) for the HeLa A1 protein. This truncated protein, termed P24*, lacked the COOH-terminal 124 amino acids of A1 and was analogous to UP1 isolated from calf thymus. During ssDNA-cellulose chromatography, P24* behaved identically to the 24kDa species described above. Finally, similar results were also obtained with the recombinant UP1, prepared with a pRC-23 expression vector in a manner similar to A1 (19). These results indicated that A1 binds much tighter to ssDNA than truncated species lacking either all or part of the COOH-terminal domain.

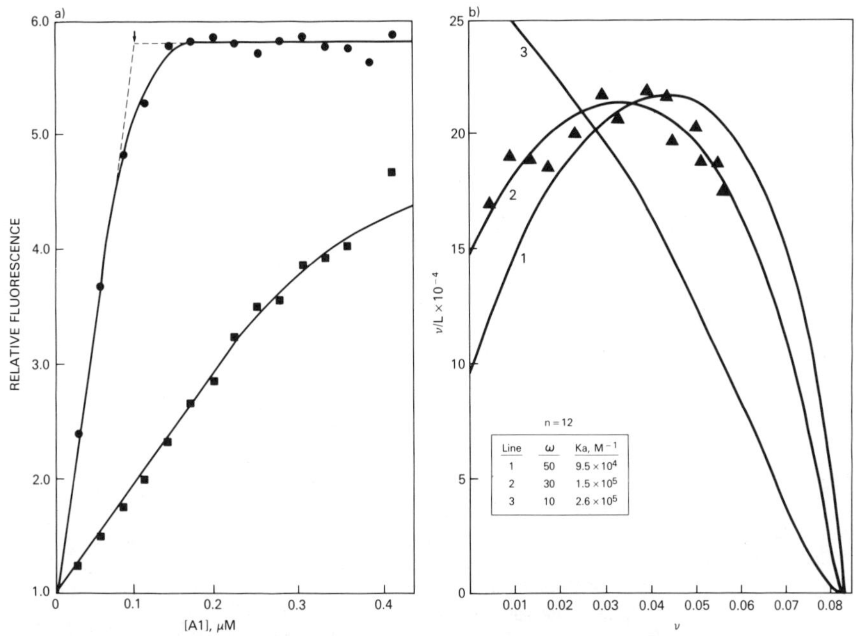

Figure 13: Fluorescence enhancement of poly(A) by recombinant A1. Experiments were performed with poly(A) at 1.25 μM (nucleotide). (A) Results obtained in the presence of 0.01 M NaCl (●—●) or 0.4 M NaCl (■—■). The *dashed line* and *arrow* illustrate estimation of *n*, the number of residues covered per protein monomer. This value was 12, since 0.1 μM protein saturated 1.2 μM residues. (B) Modified Scatchard plot of data (▲) obtained in the presence of 0.4 M NaCl. Three computer-derived theoretical curves are shown, each with *n* fixed at 12. The cooperativity parameter (ω) and intrinsic association constant (*K*) were varied as shown. A curve assuming $K = 1.5 \times 10^6\ M^{-1}$ and $\omega = 30$ fit the data points closely. (19)

Quantitative Aspects of Nucleic Acid Binding

A1-nucleic acid binding was evaluated spectrofluorimetrically using the fluorescent reporter poly(ethenoadenylate), termed poly(A) here. Binding to proteins enhances poly(A) fluorescence such that the amount of fluorescence increase is directly proportional to the amount of protein bound. Hence, titration curves of A1 with a fixed level of poly(A) can reveal equilibrium concentrations in a binding mixture of both free A1 and A1-poly(A) complex.

Under conditions of low ionic strength (0.01 M NaCl), the overall association constant is sufficiently high that all of the A1 added to the mixture is complexed with poly(A) at subsaturating ratios of protein to polynucleotide (Figure 13). Saturation of the polynucleotide with protein corresponded to 12 nucleotide residues per protein monomer (*n*). In order to determine the intrin-

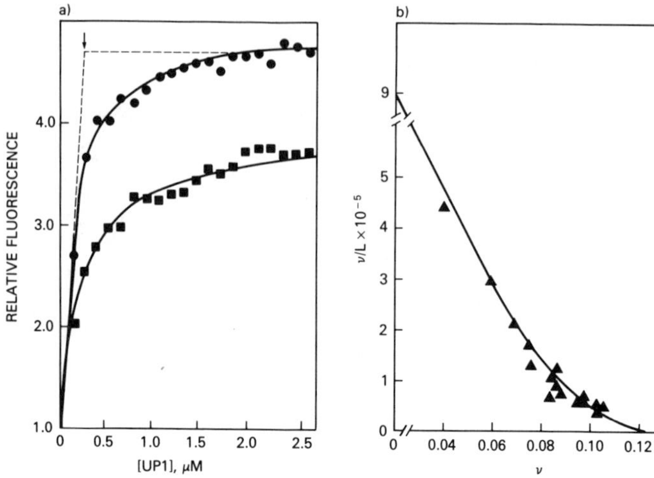

Figure 14: Fluorescence enhancement of poly(εA) by calf thymus UP1. Experiments were conducted with poly(εA) at 1.26 μM (nucleotide). (A) Results obtained in the presence of 0.01 M NaCl (●—●) or 0.05 M NaCl (■—■) The *dashed line* and *arrow* illustrate estimation of *n*, which in this case was 5. The average value of *n* from five experiments was 6.5. (B) A modified Scatchard plot of the data (▲) obtained in the presence of 0.05 M NaCl. The *solid line* is a theoretical curve assuming intrinsic association constant = 9×10^5 M^{-1}, cooperativity parameter = 1, and $n = 7$ (19).

sic association constant and degree of binding cooperativity, a titration was conducted in the presence of 0.4 M NaCl, conditions in which each binding mixture contained equilibrium concentrations of both A1-poly(A) complex and free A1. The shape of the titration curve suggested positive cooperativity; this was confirmed using a modified Scatchard analysis, as described by McGhee and von Hippel. With *n* fixed at 12, a non-linear least squares curve fitting procedure gave a best fit with the cooperativity parameter, ω, of 30 and the intrinsic association constant, *K*, of 1.45×10^5 M^{-1}. For comparison, two other curves are shown where *n* was fixed at 12, ω was fixed at 10 or 50, and *K* was varied for best fit (Figure 13b). The value of ω was examined with two curves (not shown) where *n* and *K* were fixed at 12 and 1.45×10^5 M^{-1}, respectively, and ω was either 20 or 40. These curves have roughly the same overall shape as curves 2 and 3 in Figure 12b but do not fit the data.

A series of similar poly(A) binding experiments with UP1 from calf thymus were conducted (Figure 14). In 0.01 M NaCl, saturation of the polynucleotide with UP1 corresponded to between five and seven nucleotide residues per protein monomer. Curve fitting of data obtained in the presence of 0.05 M NaCl with *n* fixed at 7 gave best fit with *K* of 9×10^5 M^{-1} and a

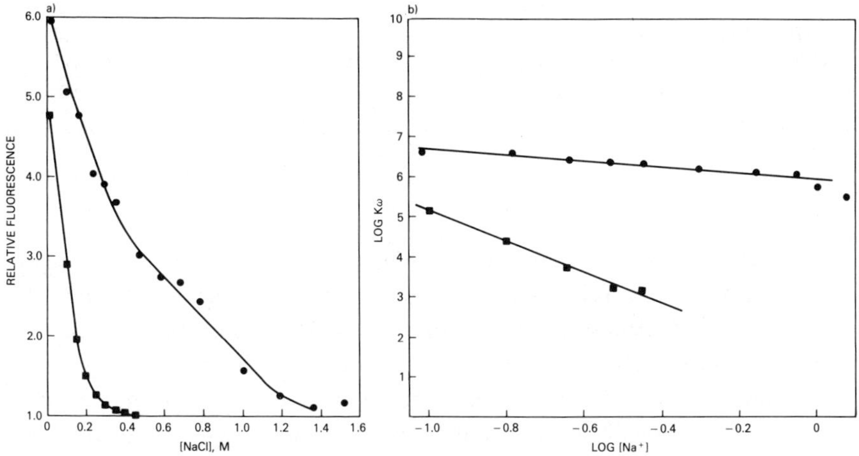

Figure 15: Reversal of protein-poly(A) binding by NaCl. For the A1-poly(A) data (●—●) prior to the addition of NaCl, A1 = 0.4 μM, [NaCl] = 0.01 M; for the UP1-poly(A) (■—■) data prior to the addition of NaCl, UP1 = 2.6 μM, [NaCl] = 0.01 M. (B) The binding levels were calculated from the amount of fluorescence enhancement assuming n = 12, ω = 30 for A1 and n = 7, ω = 1 for UP1 at all levels of NaCl. K is the intrinsic association constant (19).

cooperativity parameter of 1. Thus, this NH_2-terminal domain protein does not exhibit cooperativity.

The original level of fluorescence with A1 and UP1 could be restored by adding NaCl (Figure 15); the points of one-half reversal were ~0.4 M and ~0.08 M NaCl, respectively, for A1 and UP1. The extent of binding at each NaCl concentration along the NaCl reversal curve can be used to calculate the overall association constant, K (4,19). A plot of logK *vs.* log[Na^+] was linear for both A1 and UP1 (Figure 15b). The slopes of the plots for A1 and UP1 were ~1 and ~4, respectively, illustrating that A1 binding was much less sensitive to salt reversal than that of UP1. In additional experiments, P24* was examined, and the results obtained were virtually identical to those with the calf thymus UP1 shown in Figures 14 and 15. The results of these poly(A) binding comparisons are summarized in Table 4. In the presence of 0.4 M NaCl, A1 binding was about 10,000 times stronger than UP1 binding. This difference was not fully accounted for by the cooperativity parameter, suggesting that the COOH-terminal domain itself may contribute to the intrinsic association constant of A1.

Table 4: Summary of binding parameters with poly(A) in 0.4 M NaC1				
Protein	K (M^{-1})	ω	n	$K\omega$ (M^{-1})
A1	1.5×10^5	30	12	5×10^6
NH_2-terminal domain	2×10^3	1	7	2×10^3
COOH-terminal domain	5×10^5	10-20	1-2	5×10^6
Synthetic COOH-terminal peptide	2×10^5	25	8	5×10^6

Cooperative Binding to ssDNA

To further document cooperative binding by A1, we conducted gel retardation analysis using fdDNA as the single-stranded nucleic acid ligand. The procedure was first described by Sapp *et al.* (22). In this analysis, binding mixtures are incubated at 25°C and then subjected to agarose gel electrophoresis, where nucleoprotein complexes migrate slower than free fdDNA. Typical results are shown in Figure 16a. About 0.4 nmol of A1 (lane 7) was sufficient to complex all of the fdDNA in the mixture (3.2 nmol nucleotide residues), and the nucleoprotein complexes in this mixture migrated in a relatively sharp band. This complex probably corresponded to fdDNA molecules fully saturated with A1. Protease digestion of the complex released free fdDNA molecules with the same mobility as normal fdDNA. With binding mixtures containing less A1, fdDNA migrated either as a full nucleoprotein complex or as free molecules; the complex at each protein level migrated to approximately the same position as observed with the saturating level of protein. Thus, even though the protein amount was insufficient to fully saturate all of the fdDNA in the binding mixture, those complexes formed had about the same number of A1 monomers per DNA molecules. These results agree well with the cooperativity parameter and estimate of nucleotides covered (n=12) obtained from the poly(A) binding studies.

The fdDNA gel retardation pattern observed with recombinant UP1 (Figure 16b) was different from that with A1. At each level of UP1, all of the fdDNA was complexed, and the extent of retardation of the nucleoprotein complex was proportional to the amount of protein in the binding mixture. This is consistent with noncooperative binding and concurs with the results on UP1 binding to poly(A) (Figure 14).

Synthetic Analogue of the COOH-terminal Domain

Our results indicate that the presence of the COOH-terminal domain in A1 causes the protein to bind to poly(A) or fdDNA with higher affinity and positive cooperativity. To gain insight into the mechanism of the COOH-ter-

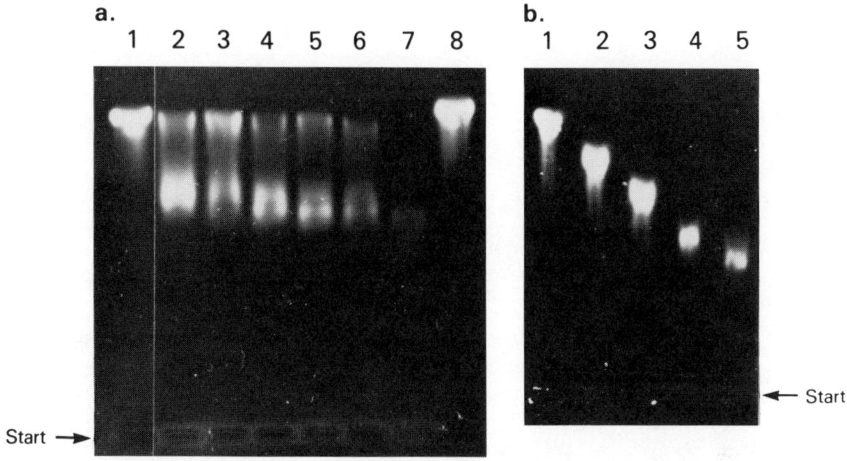

Figure 16: Gel retardation analysis of binding between fdDNA and purified recombinant proteins A1 and UP1. Photographs of ethidium bromide-stained gels are shown. (A) Nucleotide (3.2 nmol) of fdDNA and A1 for *lanes 1* through *8* was 0, 0.21, 0.17, 0.25, 0.29, 0.34, 0.42, and 0.42 nmol, respectively. For *lane 8*, proteinase K was added to a final concentration of 0.1 mg/ml, and the incubation was continued for 15 min at 35°C. (B) Binding mixtures contained 1.6 nmol (nucleotide) of fdDNA, and recombinant UP1 for *lanes 1-5* was 0, 0.034, 0.13, and 0.27 nmol, respectively (19).

minal domain effect, we prepared a synthetic peptide resembling the native domain and studied its binding to poly(A). The 124-amino acid native COOH-terminal domain is composed of repeated units of about eight amino acids with consensus sequence $\mathrm{GN}^{\mathrm{F}}_{\mathrm{Y}}\mathrm{GG}^{\mathrm{S}}_{\mathrm{G}}\mathrm{RG}$ (Table 1). We prepared a 16-residue synthetic oligopeptide representing two of the repeats, GNFGGGRGGNYG-GRG. The oligopeptide then was polymerized, and the products were size selected to obtain a higher molecular weight fraction, about 80% of which had M_r = 12,000. Hence, this synthetic polypeptide was approximately the same size as the native COOH-terminal domain and had about the same repeat interval of aromatic residues and arginine residues. Poly(A) binding analysis with the synthetic polypeptide revealed strong fluorescence enhancement to the same ΔF saturation level seen with A1 (Figure 17). One-half reversal by NaCl was not observed until concentrations >0.6 M NaCl were reached, indicating that binding is critically dependent upon non-ionic interactions. The binding site size for each protein monomer is eight nucleotides, and binding is positively cooperative, as with A1 itself. A competition assay for

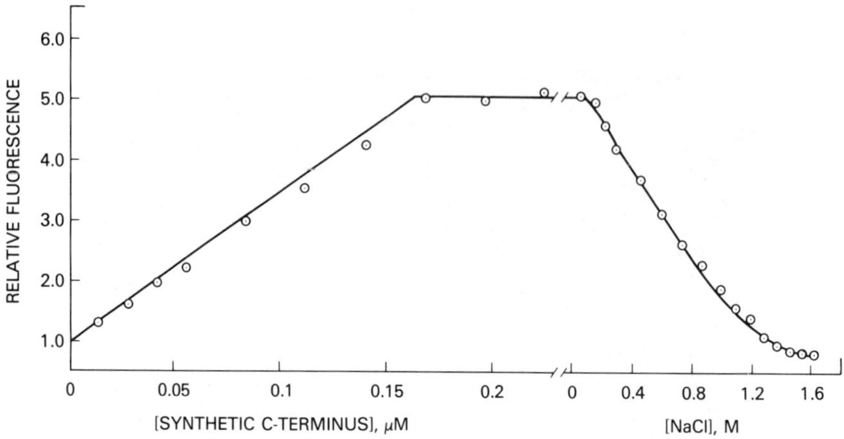

Figure 17: Fluorescence enhancement of poly(A) by the COOH-terminal domain synthetic peptide and reversal by NaCl. Experiments were conducted in the presence of 0.01 M NaCl as described in Figure 13 (19).

poly(A) binding was used to study interaction of the synthetic polymer with other nucleic acids. The results revealed that there is no specificity for binding single- *vs.* double-stranded and ribo- *vs.* deoxyribonucleic acids. The 16-residue repeat sequence monomer has no significant affinity. These results suggested that the A1 COOH-terminal domain itself could bind to nucleic acid but that it may have no single- *vs.* double-stranded specificity.

Isolation and Properties of the COOH-terminal Domain

Taken together, the comparisons of intact A1 with the NH_2-terminal domain peptides reveal that the COOH-terminal domain strongly influences nucleic acid binding by A1. The mechanism of this effect involves non-electrostatic interactions and positive cooperativity, but even without the cooperativity parameter, A1 binding to poly(A) is at least 100-fold stronger than UP1 binding. This could be due to direct interaction between the COOH-terminal domain and the polynucleotide or to an allosteric effect of the COOH-terminal domain upon binding by the NH_2-terminal domain. Studies using the synthetic COOH-terminal domain analog demonstrate that a peptide with features similar to those of the native domain is indeed capable of very strong nucleic acid binding, and structural modeling of the COOH-terminal domain of A1 indicates that the domain is capable of interfacing directly with nucleic acids. This modeling was based upon theoretical considerations of protein secondary structure, the regular spacing of most of the phenylalanine or tyrosine residues, and hydrophobic interaction of these residues with nucleic acid base residues. Therefore, we suggested that, in addition to providing protein-protein contact, the mechanism of the A1 binding involves direct interaction of the COOH-terminal domain and the nucleic acid. To test this idea, we developed conditions for production of the COOH-terminal domain using controlled proteolysis of the A1 protein (Figure 18).

The purified COOH-terminal domain of M_r 12,000 was found to bind to a ssDNA-cellulose column in the presence of 150 mM NaCl. Thus, the peptide is indeed capable of nucleic acid binding. Elution from the column was at 0.2 M NaCl, a concentration lower than that for elution of the NH_2-terminal domain (~0.35 M). Some preliminary equilibrium binding parameters of the COOH-terminal domain determined in poly(A) studies are summarized in Table 4 (35). It is interesting to note from these numbers that the binding parameters for each domain are not pecisely additive in the intact A1 protein. This suggests that the two domains are not functioning independently during nucleic acid binding. Another interesting point is that the synthetic COOH-terminal peptide binds as strongly as the actual COOH-terminal domain.

Homology of the COOH-terminal Domain with Keratin

As noted in Figure 5, there is homology at the nucleotide level between the COOH-terminal domain region of the A1 mRNA and mRNAs for various proteins of the keratin family. For example, excluding obvious deletions and insertions, the A1 cDNA between residues 600 and 730 is 71% homologous with epidermal keratin (8). Homology extends to the protein also, as a routine computer search of the data base against the 124-amino acid COOH-terminal domain shows significant match with 20 keratin proteins.

Amino acid alignment between the COOH-terminal domain and a portion of the 68kDa, type II cytoskeletal keratin is shown in Figure 19. The alignment corresponds to the first 99 residues of the COOH-terminal domain. There is 37% match, and differences are conservative. We believe this homology with keratin is interesting because it suggests the idea that A1 could complex with cytoskeletal components and thus contribute to the macromolecular organization of the hnRNP complex.

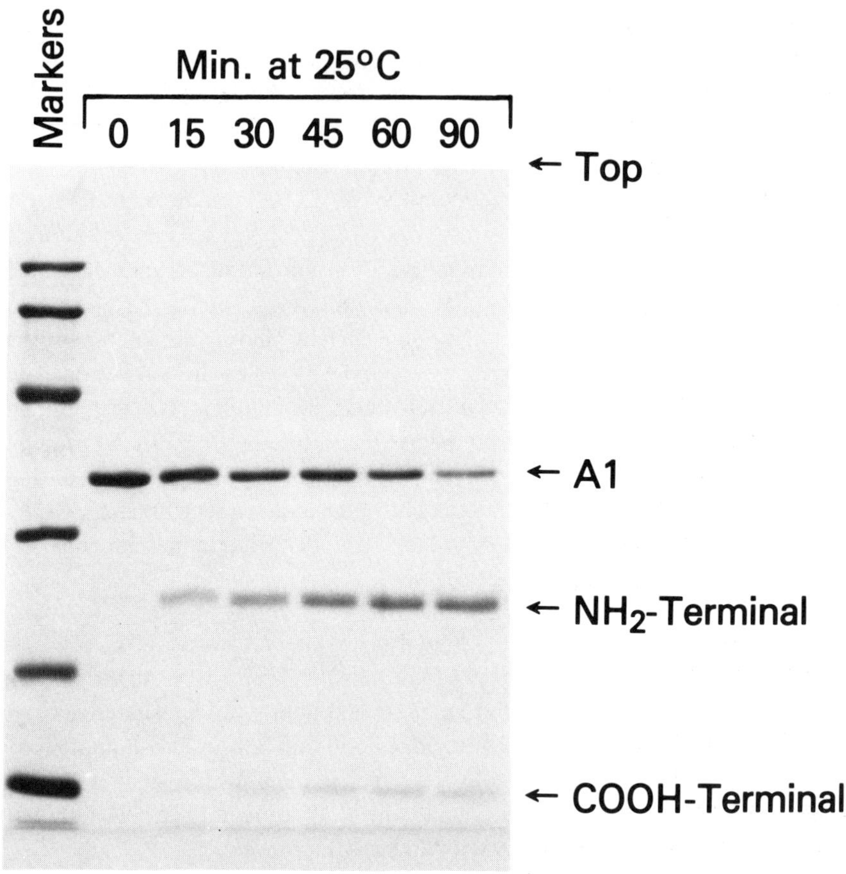

Figure 18: SDS-polyacrylamide gel analysis showing results of controlled proteolysis of A1 with trypsin. Digestion was at 25°C with an A1/trypsin ratio of 100. The gel concentration was 12.5%, and the time of digestion in minutes is shown. The gel was stained with Coomassie blue.

The NH_2-terminal Domain

Isolation of the Two 92-Amino Acid Domains (16)

To determine if each of the two 92-amino acid sequence repeats in the NH_2-terminal domain represents an independently folded, globular domain, we used *S. aureus* V-8 protease to cleave the protein after two glutamic acid

```
                          10        20        30        40        50
A1—               SGSGNFGGGRGGGFGGNDNFGRGGNFSGRGGFGGSRGGGGYGGSGDGYNGFGND-GSNFG
Mouse Keratin—    VSSSSTTSASAGGFGGGYGGGVGVGGGARSGFGGGSGFGG-GSGISGSSGFGGGSGSGFG
                         70        80        90       100       110       120

                 60        70        80        90       100
A1—               GGGSYNDFGNYNNQSSNFGPMKGGNFGGRSSGPYGGGGQYF
Mouse Keratin—    GGSGFSGSSGFGGGSSGFGSGSGGRSGVSGGGLSSGSSRGG
                         130       140       150       160


                          10        20        30        40        50
A1—               SGSGNFGGGRGGGFG-GNDNFGRGGNFSGRGGFGGSRGGGGYGGSGDGYNGFGNDGSNFG
Human Keratin—    SGGGSRGGG-GGGYGSGGSSYGSGGCSYGSGG-GGGGGRGSYGSGGSSYGSGGSSYGSGG
                 360       370       380       390       400       410

                 60        70        80        90       100       110
A1—               GGGSYNDFGNYNNQSSNFGPMKGGNFGGRSSGPYGGGGQYFAKPRNQGGYGGSSSSSSYG
Human Keratin—    GGGGHGSYGS-GSSSGGYRGGSGGGGGGSSGGRGSGGGSSGGSSGGRGSSSGGVKSSGGS
                  420       430       440       450       460       470

                 120
A1—               SGRRF
Human Keratin—    SSVRF
```

Figure 19: Example of sequence homology between the 124-residue COOH-terminal domain of A1 and keratin proteins, rodent, and human 68kDa type II cytoskeletal, component I keratin. Alignment was by the ZIFASTA program of D. J. Lipman, NIH. The GeneBank Data Bank as of 7/1/88 was evaluated. The homology score for A1 and keratin was >22 standard deviations above mean. Sequences not shown did not exhibit significant homology. Numbering of residues was from the NH_2 end of the 124-residue COOH-terminal domain of A1 (*top*) or the NH_2-terminal residue of keratin (*bottom*).

residues (numbers 92 and 184) that lie almost exactly at the COOH-terminal ends of each of the 92-residue sequence repeats. *S. aureus* V-8 protease first cleaves after glutamic acid 184, removing 11 amino acids from the COOH-terminal. More extensive digestion results in cleavage at glutamic acid 92, producing two fragments in approximately equal yield that differ slightly in apparent size. (The COOH-terminal portion of A1 is completely digested to small peptides under these conditions.) When a mixture of these two fragments is chromatographed on a ssDNA cellulose column, the fragment with slower mobility upon SDS-polyacrylamide gel electrophoresis is slightly

SEQUENCES FOUND

14 K L F I G G L S F E T T D E S L R	60% OF X-LINKED PEPTIDE
105 K I F V G G I K E D T E E H H L	12% OF X-LINKED PEPTIDE
146 G F A F V T F D D H D S V D K I V I	8% OF X-LINKED PEPTIDE
55 G F G F V T Y A T V E E V D A A M N	7% OF X-LINKED PEPTIDE

Figure 20: Summary of sequence analysis and the amount of cross-linking to each major, labeled tryptic peptide after UV irradiation of the A1-[^{32}P]p(dT)8 complex.

retained, whereas the other fragment is not. Amino acid analyses and sequencing indicated that the "larger" fragment corresponds to residues 1-92 and the "smaller" fragment corresponds to residues 93-184.

Identification of the Tryptic Peptide Cross-linked to [^{32}P]p(dT)8 (16)

A1 protein was mixed with [^{32}P]p(dT)$_8$, irradiated with ultraviolet light, and digested with trypsin. Radioactive peptides were isolated with a combination of anion-exchange and ion-pairing HPLC. The isolated peptides were identified by gas-phase sequence analysis. Although sequencing revealed that a mixture of peptides was present, the peptides could be assigned relative to the known A1 sequence (Figure 20). It was also possible to obtain the relative amounts of each peptide based on sequencing yields. These results were combined with radioactivity quantification to obtain an estimate for the amounts of each peptide present. The ratio of peptide yield obtained to the amount of peptide expected was calculated as a precaution to ensure that peptides sequenced were not noncross-linked contaminants. We conclude that cross-linking of A1 to [^{32}P]p(dT)$_8$ resulted in covalent bonds at four phenylalanine residues, phenylalanines 16, 58, 107, and 149 (Figure 20). It is unlikely that the multiple cross-linking sites are nonspecific, since cross-linking is inhibited by either a concentration of NaCl known to disrupt the A1-p(dT)$_8$ complex or by an excess of p(dA)$_8$. In addition, the phenylalanines involved are in exactly analogous positions in the 92-amino acid fragments of the NH_2-terminal domain, and both of these fragments can independently bind to p(dT)$_8$. Thus, phenylalanines 16 and 107 occur in the same position of a highly conserved region of each fragment with the sequence Lys-Leu/Ile-Phe-Ile/Val-Gly-Gly-Leu/Ile, *i.e.*, "RNP consensus sequence 2." Phenylalanines 58 and 149 correspond to the second phenylalanine in the "RNP consensus sequence 1" of Adam *et al.* (18), Arg-Gly-Phe-Gly/Ala-Phe-Val-Thr-Tyr/Phe (see Figure 9).

Our finding that at least four phenylalanines in A1 are at the protein-nucleic acid interface, where they may form part of an RNA-binding pocket, supports the general idea that hydrophobic interaction between aromatic amino acids and nucleotide bases represents a general mechanism by which numerous proteins recognize and bind to single-stranded nucleic acids (33). The observation that aromatic residues in the COOH-terminal domain were not cross-linked to p(dT)$_8$ probably reflects a difference in the detailed mechanism by which the two A1 domains interact with polynucleotide bases. It should be noted in this context that binding by fd gene 5 protein involves interaction between phenylalanine and a base residue (34), but no ultraviolet cross-linking is detected with this residue.

The internal sequence repeats in the NH_2-terminal domain apparently do represent two nucleic acid-binding moieties, since we found that the two purified V-8 protease fragments, containing residues 1-92 and 93-184, respectively, can be independently cross-linked to [^{32}P]p(dT)$_8$. The extent of cross-linking of the two fragments, as judged by the relative intensities of the cross-linked bands on an SDS-polyacrylamide gel, was greater for the 1-92 fragment than for the 93-184 fragment. This is consistent with the behavior of these two polypeptides on a ssDNA-cellulose column, since the 1-92 fragment has greater affinity for this column than does the 93-184 fragment. The results indicate that although the two repeats in the NH_2-terminal domain share some localized sequence homology, they are not equivalent in binding.

CONCLUSIONS

With the aim of understanding structure-function relationships of a eukaryotic nucleic acid-binding protein, we have studied the A1 protein of the nuclear hnRNP complex. A1 is a highly conserved two-domain protein. Both its NH_2- and COOH-terminal domains possess nucleic acid-binding pockets, and the COOH-terminal domain also confers positive cooperativity through protein-protein interaction with the COOH-terminal domain of an adjacent A1 molecule. A diagram illustrating these points is shown in Figure 21. The COOH-terminal domain is shown binding in an elongated fashion to the back

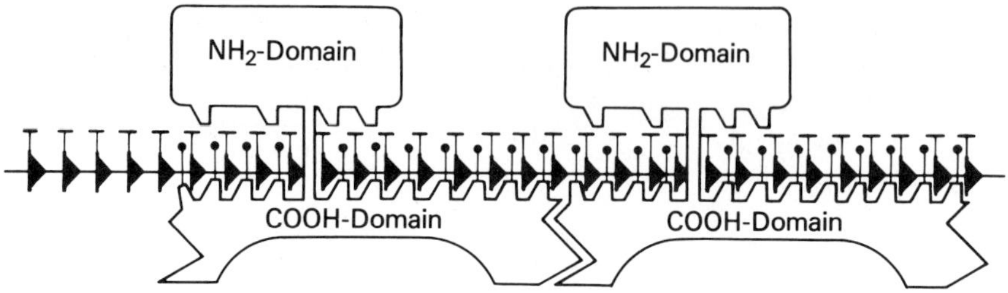

Figure 21:

side of the polynucleotide chain spanning 12 nucleotide residues. The NH_2-terminal domain, on the other hand, loops around and binds to the front or base side of the nucleic acid, spanning seven nucleotide residues; the NH_2-terminal domain confers specificity for single-stranded (*vs.* double-stranded) polynucleotides by virtue of this type of binding. COOH-terminal aromatic residues make contact with base residues as illustrated, yet the site size found for the COOH-terminal fragment (1-2 residues) is not accomodated in the model.

ACKNOWLEDGMENTS

I thank my collaborators, Kenneth Williams, Barbara Merrill, and Kathy Stone at Yale University and Rich Karpel at University of Maryland and my associates, especially Fabio Cobianchi, Amal Kumar and Al Fornace.

REFERENCE

1. Herrick, G., and Alberts, B. (1976). *J. Biol. Chem.*, **251**:2124-2132.
2. Planck, S. R., and Wilson, S. H. (1980). *J. Biol. Chem.*, **255**:11547-11556.
3. Herrick, G., and Alberts, B. (1976). *J. Biol. Chem.*, **251**:2133-2141.
4. Karpel, R. L., and Burchard, A. C. (1980). *Biochemistry*, **19**:4674-4682.
5. Karpel, R. L., Miller, N. S., and Fresco, J. R. (1982). *Biochemistry*, **21**:2102-2108.
6. Valentini, O., Biamonti, G., Mastromei, G., and Riva, S. (1984). *Biochim. Biophys. Acta*, **782**:147-155.
7. Valentini, O., Biamonti, G., Pondolfo, M., Morandi, C., and Riva, S. (1985). *Nucleic Acids Res.*, **13**:337-346.
8. Cobianchi, F., SenGupta, D. N., Zmudzka, B. Z., and Wilson, S. H. (1986). *J. Biol. Chem.*, **261**:3536-3543.
9. Merrill, B. M., LoPresti, M. B., Stone, K. L., and Williams, K. R. (1986). *J. Biol. Chem.*, **261**:878-883.

10. Williams, K. R., Stone, K. L., LoPresti, M., Merrill, B. M., and Planck, S. R. (1985). *Proc. Natl. Acad. Sci. USA*, **82**:5666-5670.
11. Mayrand, S., Setyono, B., Greenberg, J. R., and Pederson, T. (1981). *J. Cell Biol.*, **90**:380-384.
12. Beyer, A. L., Christensen, M. E., Walker, B. W., and LeStourgeon, W. M. (1977). *Cell*, **11**:127-138.
13. Wilk, H. E., Werr, H., Friedrich, D., Kiltz, H. H., and Schafer, K. P. (1985). *Eur. J. Biochem.*, **146**:71-81.
14. Leser, G., and Martin, T. E. (1981). **In**: *DNA: Protein Interaction and Gene Regulation*, (Thompson, E. B., and Papaconstantinou, J., Eds.) pp. 97-106, University of Texas Press, Austin, TX.
15. Wilson, S. H., Cobianchi, F., and Guy, H. R. (1981). **In**: *DNA: Protein Interaction and Gene Regulation*, (Thompson, E. B., and Papaconstantinou, J., Eds.) pp. 129-146, University of Texas Press, Austin, TX.
16. Merrill, B. M., Stone, K. L., Cobianchi, F., Wilson, S. H., and Williams, K. R. (1988). *J. Biol. Chem.*, **263**:3307-3313.
17. Swanson, M. W., Nakagawa, T. Y., LeVan, K., and Dreyfuss, G. (1987). *Mol. Cell. Biol.*, **7**:1731-1739.
18. Adam, S. A., Nakagawa, T., Swanson, M. S., Woodruff, T. K., and Dreyfuss, G. (1986). *Mol. Cell. Biol.*, **6**:2932-2943.
19. Cobianchi, F., Karpel, R. L., Williams, K. R., Notario, V., and Wilson, S. H. (1988). *J. Biol. Chem.*, **263**:1063-1071.
20. Crowl, R., Seamans, G., Lomedico, P., and McAndrew, S. (1985). *Gene (Amst.)*, **38**:31-38.
21. Kumar, A., Williams, K. R., and Szer, W. (1986). *J. Biol. Chem.*, **261**:11266-11273.
22. Sapp, M., Knippers, R., and Richter, A. (1986). *Nucleic Acids Res.*, **14**:6803-6820.
23. Corden, J. L., Cadena, D. L., Ahearn, J. M., Jr., and Dahmus, M. E. (1985). *Proc. Natl. Acad. Sci. USA*, **82**:7934-7938.
24. Allison, L. A., Moyle, M., Shales, M., and Ingles, C. J. (1985). *Cell*, **42**:599-610.
25. Detera, S. D., Becerra, S. P., Swack, J. A., and Wilson, S. H. (1981). *J. Biol. Chem.*, **256**:6933-6943.
26. Lahiri, D. K., and Thomas, J. O. (1986). *Nucleic Acids Res.*, **14**:4077-4094.
27. Lapeyre, B., Bourbon, H., and Amalric, F. (1987). *Proc. Natl. Acad. Sci. USA* **84**:1472-1476.
28. Theissen, H., Etzerodt, M., Reuter, R., Schneider, C., Lottspeich, F., Argos, P., Luhrmann, R., and Philipson, L. (1986). *EMBO J.*, **5**:3209-3217.
29. Habets, W. J., Sillekens, P. T. G., Hoet, M. H., Schalken, J. A., Roebroek, A. J. M., Leunissen, J. A. M., Van de Ven, W. J. M., and van Venrooij, W. J. (1987). *Proc. Natl. Acad. Sci. USA*, **84**:2421-2425.
30. Haynes, S. R., Rebbert, M. L., Mozer, B. A., Fouquignon, F., and Dawid, I. B. (1987). *Proc. Natl. Acad. Sci. USA*, **84**:1819-1823.
31. Sachs, A. B., Bond, M. W., and Kornberg, R. D. (1986). *Cell*, **45**:827-835.
32. Sachs, A. B., Davis, R. W., and Kornberg, R. D. (1987). *Mol. Cell. Biol.*, **7**:3268-3276.
33. Coleman, J. E., Williams, K. R., King, G. C., Prigodich, R. V., Shamoo, Y., and Konigsberg, W. H. (1986). *J. Cell. Biochem.*, **32**:305-326.
34. Brayer, G. D., and McPherson, A. (1983). *J. Mol. Biol.*, **169**:565-596.
35. Kumar A., Casas-Finet, J. R., Luneau, C. J., Karpel, R. L., Merrill, B. M., Williams, K. R. and Wilson, S. H. (1990). *J. Biol. Chem.* in press.

5

TRANSCRIPTION OF THE PRE-rRNA GENES OF EUKARYOTES

R. V. Quincey[1]

The University of Birmingham
School of Biochemistry
P.O. Box 363
Birmingham, England B15 2TT

1 Deceased September 1990

INTRODUCTION

This chapter is concerned with the way cloned pre-rRNA genes have been exploited to determine which DNA sequences act as signals for the initiation of transcription and to learn how these signals are recognized and how expression is controlled. The discussion will be restricted to nuclear pre-rRNA genes coding for the rRNA of cytosolic ribosomes and consequently will exclude discussion of the pre-rRNA genes of cell organelles. The chapter begins with an outline of the structure of pre-rRNA genes and of the processing of pre-rRNA to yield mature rRNA species.

PRE-rRNA GENES AND THE PROCESSING OF PRE-rRNA

Eukaryotes contain one hundred to one thousand copies of the pre-rRNA gene that encodes the 25-28S rRNA, 5.8S rRNA and 17-18S rRNA of ribosomes. The structure of these genes was reviewed by (1) Long and David (1980). Generally, the pre-rRNA genes are organized as tandem arrays of direct repeats separated from each other by several kb of nontranscribed spacer DNA, although the term, non-transcribed spacer is a misnomer because in some species much of this region is transcribed (2,3,4). In the cell, rDNA is incorporated into nucleoli where pre-rRNA biosynthesis, processing and assembly into ribosomes takes place. Assembly requires the transport into the nucleolus of ribosomal proteins and of 5S rRNA, since 5S rRNA genes are usually located separately from pre-rRNA genes within genomes and are not part of nucleoli. This organization of genes from rRNA is not invariant. In *Dictyostelium discoides, Physarum polycephalum* and *Tetrahymena pyriformis*, pre-RNA genes are found as extrachromosomal, dimeric, linear palindromes. In *D. discoides* and *Saccharomyces cerevisiae*, genes for 5S rRNA are in the nontranscribed spacer.

Pre-rRNA genes have a common structure with a large external transcribed spacer at the 5′ end followed by the coding regions for 17-18S rRNA, 5.8S rRNA and 25-28S rRNA in that order, separated from each other by short internal transcribed spacers. Pre-rRNA genes differ significantly in size between organisms and are largest in higher eukaryotes. Variations are due mainly to differences in the sizes of the external transcribed spacer and to a lesser extent to differences in the mature rRNA coding sequences, particularly those for 25-28S rRNA. The size of the nontranscribed spacer varies considerably between species.

Introns within the 28S rRNA coding region of some copies of the pre-rRNA gene are common in Dipteran insects (5). In *D. melanogaster*, introns of approximately 5 kb, or of 3.5 kb, occur in about half of the pre-rRNA genes. No mature rRNA originates from these intron^{+} copies of the pre-rRNA gene, so no splicing mechanism is required. By contrast, in some strains of *Tetrahymena* an intron of 400 bp is present in all copies of the 28S rRNA coding region. This sequence is removed from the pre-rRNA in a self-splicing reaction that precedes other processing reactions (6). The intron in the *Tetrahymena* pre-rRNA gene shares conserved sequences with a large group of introns found in genes coding for proteins and rRNA in mitochondria, and genes for rRNA in *Neurospora* and *P. polycephalum*. These introns may have a common ancestor and share the same splicing mechanism (7).

Processing of pre-rRNA involves several cleavage steps and methylation of ribose and of bases in the regions corresponding to the mature rRNA molecules (8). In insects further processing occurs: the 28S rRNA is cut into two approximately equal fragments (28Sα and 28Sβ) that hydrogen-bond to each other, and in some insects the 5.8S rRNA is cut into two (5).

The Location of Initiation Sites

The development of nuclease S1 and primer extension techniques for mapping the 5′ ends of RNA onto the corresponding DNA sequence provided a way of locating initiation sites precisely (9,10). However, these determinations have not been straightforward because new 5′ ends are generated during processing. Thus, unless it is known that the pre-rRNA preparation contains a reasonable proportion of molecules with the original 5′ end, a processing site may be mapped instead of the initiation site. Two methods have been used to estimate the proportion of pre-rRNA that retains the original 5′ ends. Both rely on detecting the structure (p)ppXp... which occurs only at original 5′ ends. In one method, RNA is labelled with ^{32}Pi *in vivo*, pre-rRNA is purified, hydrolysed and the molar proportion of (p)ppXp is determined. In the second method, pre-rRNA is reacted with [α-^{32}P]GTP and a preparation of

	-40	-30	-20	-10 → +1 → +10
Mouse		ATCTTT	TATTG	TACTGACACGC
Rat		ATCTTT	TATTG	TGCTGACACGC
Human	ATCTTT		TTTTGG	TGCTGACACGC
X. laevis		CTACGCTTTT	GTGCGGCCAGGAAGGTAGGG	
X. clivii		CTACGCGTTT	GTGCCGACAGGAAGGTAGGG	
X. borealis		CTACGCTTTT	GTGCGGACAGGAAGGTAGGG	
D. melanogaster		TTT	G	TAGGT
T. pyriformis			G	T
D. discoideum			T	ATACATATA
P. polycephalum			A	ATACATATA
S. carlsbergensis			A	AGGTACTTCATGCGAAAGC
S. rosei			G	AGGAACTTCATGCGAAAGC
A. castellanii	TTTT		GT	TATATATA

Figure 1: Major sequence homologies around the transcription initiation site of eukaryotic pre-rRNA genes. Reproduced from Sommerville, (114) with the inclusion of data for *A. castellanii* (46). Note the conserved G at -16 and T at -1 bp. The direction of transcription is indicated by an arrow. Reprinted by permission from *Nature*, **310**, 189-190. Copyright ©1984 MacMillan Magazines Ltd.

guanyltransferase, or with S-adenosyl-L-[methyl-^{3}H]methionine, GTP and a capping enzyme complex from Vaccinia virus. Both cap RNA substrates terminating in (p)ppXp, but not pXp, at the 5′ end (11,12,13). After hydrolysing the capped RNA with P1 nuclease, the identity of X in GpppX may be determined by chromatography, or electrophoresis against authentic samples of the possible forms of GpppX.

A relatively large proportion (20%) of pre-rRNA from yeast, *T. pyriformis, D. melanogaster* and *X. laevis* retains the original 5′ end (14,15,16,11). This proportion is 10-20% in pre-rRNA from mice, rats and humans (17,12,13). In *D. discoides*, pre-rRNA is processed rapidly to a pre-17S rRNA, most of which retains the original 5′ end (18). In these organisms, and others, the sites of initiation of transcription of pre-rRNA genes have been mapped. There is little sequence conservation around the initiation sites, except in closely related species (Figure 1). Evidence that these are initiation sites rather than processing sites comes from 1, observations that these are the most upstream sites that can be mapped, 2, the identity of the predicted 5′ nucleotide with that of pre-rRNA, and 3, the correspondence of these sites with those located in transcription assays.

Termination Sites

Efforts to locate termination sites were also hampered by the rapid processing of pre-rRNA. In some cases, sites close to the 3′ ends of the 28S rRNA coding region that were regarded as termination sites, have recently been identified as sites of rapid processing.

In the mouse and the rat, transcription terminates about 565 bp downstream of the 28S rRNA coding region, in a pyrimidine rich sequence just upstream from several copies of an 18 bp element (the *Sal* box) that is almost identical in the two organisms (19,20). In the mouse, termination requires a protein factor that binds to the *Sal* box and is accompanied by processing to remove eleven nucleotides from the 3′ end of the primary transcript (21). By contrast, transcription of the *X. laevis* gene proceeds right through the nontranscribed spacer to a termination site 213 bp upstream from the next gene (3,2). Termination at this site requires a 12 bp element which includes a 7 bp sequence that is an essential part of the processing site several kb upstream that yields 40S pre-rRNA. Surprisingly, mutation of the 7 bp sequence in the terminator abolishes this processing reaction (22). In *Drosophila* transcription also proceeds well into the nontranscribed spacer, but does not terminate at a specific site (4). Transcription in *S. cerevisiae* terminates 210 bp downstream from the 25S rRNA coding region, but there is also fail-safe terminator approximately 450 bp further downstream (23).

TRANSCRIPTION OF PRE-rRNA GENES

RNA Polymerase I

Pre-rRNA genes are transcribed by RNA polymerase I. Evidence for this has been obtained most easily in higher eukaryotes where RNA polymerase I activity and pre-rRNA biosynthesis are insensitive to α-amanitin whilst the activity of RNA polymerases II and III and the biosynthesis of pre-mRNA, 5S rRNA and tRNA are sensitive to this toxin (24,25,26,27). RNA polymerase I enzymes have been isolated from many eukaryotic organisms. These enzymes contain two large subunits with molecular weights of about 190,000 and 135,000 and several smaller subunits, some of which are shared with RNA polymerases II and III (28). Early attempts to show that RNA polymerase I could transcribe pre-rRNA genes specifically *in vitro* relied on showing by hybridization experiments that the total transcript was enriched for pre-rRNA sequences, and some successes were claimed (29,30). However, to prove that transcription is specific it is necessary to show that the transcript is initiated *in vitro* at the same site as *in vivo*. As techniques for locating initiation sites became available, it became clear that RNA polymerase I by itself was incapable of specifically transcribing cloned pre-rRNA genes.

Assays for the Specific Transcription of Cloned pre-rRNA Genes

The failure of purified RNA polymerase I enzymes to transcribe cloned pre-rRNA genes specifically suggested that transcription factors are required. Assays for transcription of cloned RNA genes have been developed which employ whole cells, or cell extracts. These assays are complementary: cell extracts are particularly useful for purifying transcription factors, but reflect the *in vivo* situation imperfectly.

Cell extracts consist either of an S100 supernatant obtained by spinning at 100,000 g for 60 min an homogenate of cells swollen in hypotonic buffer (31,24), or of whole cell extracts obtained after adding $(NH_4)_2SO_4$ to such an homogenate (32,33). RNA biosynthesis in both types of extract is dependent on adding DNA. The specificity of initiation and the amount of transcription may be determined in three ways. In one method, the DNA template is cut just downstream from the initiation site and the size and amount of the run-off transcript copied from the truncated DNA in the presence of ^{32}P-labelled nucleoside triphosphate is determined by autoradiography of a polyacrylamide gel of the ^{32}P-RNA. In the other methods unlabelled RNA is analysed in a nuclease S1 protection assay, or a primer extension assay, where the ^{32}P-labelled DNA hybridization probe or primer is in excess.

When whole cells are used to assay the transcription of homologous pre-rRNA genes, the problem of distinguishing between the products of the cloned gene and of the endogenous pre-rRNA genes arises. This problem was overcome initially for *X. laevis* by injecting circular recombinant DNA molecules into the nuclei of oocytes (34). Eight to eighteen hours later, the nuclei were isolated and the contents spread for electron microscopy. The small circular plasmid molecules were readily distinguished from the endogenous DNA. More than 90% of these plasmid molecules were transcriptionally inactive and displayed a beaded structure typical of chromatin. A dense gradient of nascent RNA on other plasmid molecules was morphologically indistinguishable from than on endogenous pre-rRNA genes which were being transcribed actively. However, visualization of transcription complexes by electron microscopy does not permit precise localization of the initiation site, or provide a quantitative measure of the overall extent of transcription of the cloned gene, because these transcription complexes are seen infrequently (35).

Alternative methods for detecting transcripts from whole cells use nuclease S1 protection and primer extension assays, or blotting and hybridization assays. In the original application, advantage was taken of the divergence in the sequences of the external transcribed species of the pre-rRNA genes of the interfertile species *X. laevis* and *X. borealis*. In these two organisms, the sequences required for transcription of pre-rRNA genes are conserved, so pre-rRNA genes from *X. laevis* are transcribed in *X. borealis* oocytes, but sequence divergence in the external transcribed spaces means that a hybridization probe can be used that recognizes specifically transcripts of the pre-rRNA gene of *X. laevis* (35,36). This method has also been used to measure transcription of mouse pre-rRNA genes transfected into rat cells (37) and of human pre-rRNA genes transfected into monkey cells (38). This type of strategy has limitations, however, because pre-rRNA gene transcription commonly shows a high degree of species specificity (39,26) and it is desirable to use homologous systems.

In homologous systems, the problem of discriminating against transcripts of the host genes is overcome by incorporating foreign DNA within the cloned copy and using a ^{32}P-labelled fragment incorporating the foreign DNA as a hybridization probe, or as a primer. This strategy has been used particularly for yeast (40,41,42). In the nuclease S1 protection assay, the 3′ end of the probe consists of rDNA spanning the initiation site and the 5′ end labelled with ^{32}P consists of the foreign DNA (Figure 2). In the primer extension assay, all of the primer is of foreign DNA: transcripts from yeast have also been detected on Northern blots using the foreign DNA as a hybridization probe (42,41). In each case, yeast was transformed with an *Escherichia coli* - yeast shuttle vector containing a pre-rRNA minigene that incorporated foreign DNA.

Species Specificity

The transcription of pre-rRNA genes shows a high degree of species specificity: mouse genes are not transcribed in human cell extracts (24); human and *Acanthamoeba* genes are not transcribed in mouse cell extracts (39) and *D. virilis* genes are not transcribed in extracts of cells from *D. melanogaster* (26). This specificity is not absolute. Mouse and human genes are expressed in transfected rat and monkey cells respectively (37,38) and *X. laevis* genes are expressed in *X. borealis* oocytes. *X. laevis* pre-rRNA genes are also expressed in extracts of mouse cells, but the specificity depends critically on the conditions of the assay (43). At a low salt concentration of 40

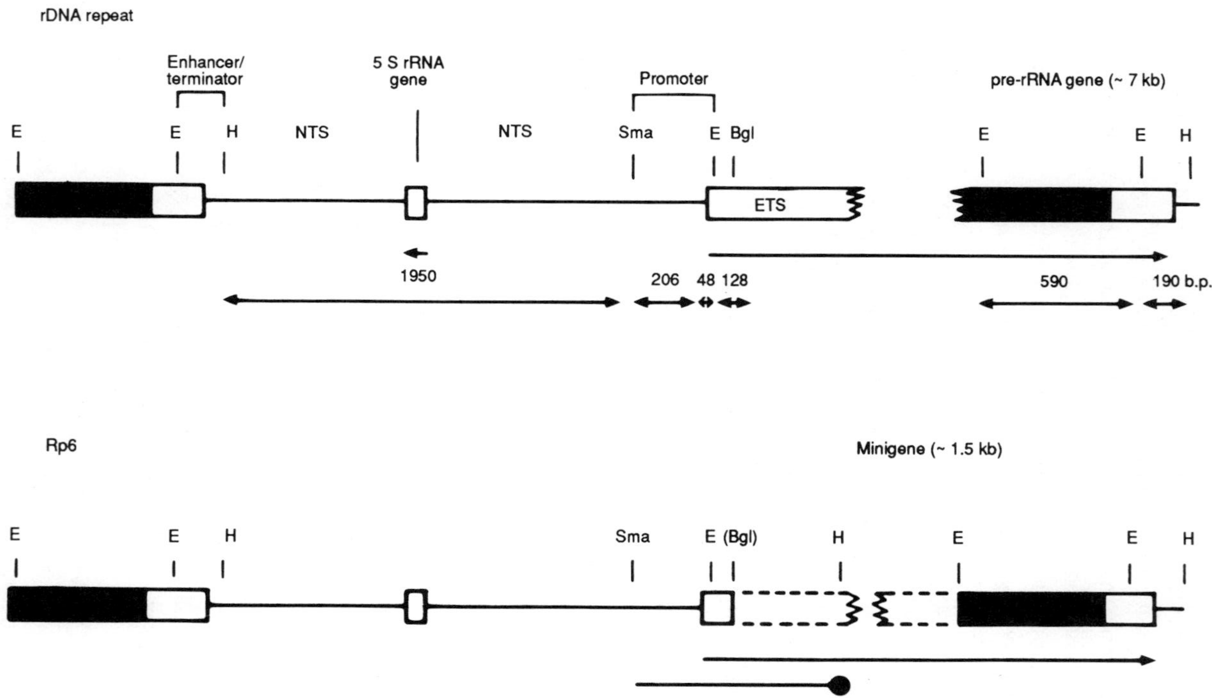

Figure 2: Structure of the rDNA repeat of *S. cerevisiae* and of the pre-rRNA minigene in Rp6 (40). Pre-rRNA and 5S rRNA gene sequences are boxed, with the 25S rRNA coding region shaded. The direction of transcription is shown by horizontal arrows. Boxes enclosed by dotted lines show sequences of vector, or foreign DNA that are transcribed within the minigene. The expected transcript and the DNA hybridization probe, labelled at the 5′ end with $^{32}P_i$ (●), used to detect the transcript in a nuclease S1 protection assay are shown.
Abbreviations used: NTS, nontranscribed spacer; ETS, external transcribed spacer. Restriction sites are shown for *Eco*R1 (E). *Hind*III (H), *Sma*I (Sma), BglII (Bgl) and Sau3A. Bgl in parentheses indicates that the site is lost.

mM KCl, the *X. laevis* gene was transcribed from the correct initiation site, but 10 times less efficiently than the mouse gene. At 90 mM KCl, optimal for transcription of the homologous mouse gene, transcription of the *X. laevis* gene was approximately 100 times greater, but was initiated at -4 bp. By contrast to pre-rRNA genes, many genes coding for proteins, 5S rRNA and tRNA are accurately transcribed in heterologous systems (28). These differences in behaviour in heterologous systems may reflect the absence of a consensus sequence in pre-rRNA genes from different species and the presence of such sequences in genes transcribed by RNA polymerases II and III.

The Role of Transcription Factors

When a pre-rRNA gene is incubated for 10-15 min at 30-37°C with a competent cell extract lacking nucleoside triphosphates, a stable preinitiation complex is formed. Evidence for this comes from the interpretation of pre-binding competition experiments in which an excess of one template is preincubated with a cell extract. If a complex forms, no factors will be left to bind to a second template added later. Thus when nucleoside triphosphates are added to start the reaction transcription is from the first template only. Once formed, such complexes are stable for at least 60 min during repeated rounds of gene transcription (44,45). RNA polymerase I is not required for forming the stable pre-initiation complex (27,46,47).

The number of transcription initiation factors required for the expression of pre-rRNA genes is unknown. Estimates of the minimum number have been obtained by reconstituting the transcriptional activity of fractionated extracts. Mishima *et al.* (48) used chromatography on phosphocellulose to separate extracts of mouse and human cells into 4 fractions (A-D), including one (C) that contained RNA polymerase I. Each was required for accurate and efficient transcription. Using a similar fractionation procedure for extracts from cells of the same organisms, (27) found that fractions A, C and D were sufficient.

To investigate how the factors in fractions A-D of mouse and human cell extracts might act, Mishima *et al.* (48) analysed the transcription of human and mouse pre-rRNA genes using heterologous combinations of fractions A-D from human and mouse cells; expression required homologous fraction D, but fractions A-C were interchangeable. Pre-binding competition experiments showed that a factor in D forms a stable pre-initiation complex with a region of the homologous pre-rRNA gene spanning -90 and -30 bp and forms no complex with a heterologous pre-rRNA gene promoter (27). By using similar competition experiments to determine the effectiveness of mutated derivatives of the promoter to bind fraction D (TIF-D), Nagamine *et al.* (49)

concluded that DNA from -40 bp to -1 bp was essential. Comparisons of the DNA sequences of the mouse and human promoters suggested that the sequence responsible for the species-specific binding of TIF-D might lie between -32 bp and -21 bp as the sequences either side are homologous (Figure 1). Accordingly, Safrany *et al.* (50) synthesized a hybrid promoter in which the mouse sequence between -32 bp and -14 bp was inserted into the human promoter and found that it was transcribed in mouse extracts and could bind mouse TIF-D, but not human TIF-D. Thus, the species specificity of transcription of pre-rRNA genes may be explained at least in part by the sequence-specific binding of TIF-D.

The factor TIF-1B isolated from mouse cells by Clos *et al.* (47) is probably identical to TIF-D. TIF-1B is present in growing and quiescent cells in comparable amounts. TIF-1B has been purified 100,000-fold and forms a stable pre-initiation complex with mouse pre-rRNA genes. The molecular weight of TIF-1B is about 80,000. The binding site for TIF-1B has been located by exonuclease III protection experiments in which complexes between TIF-1B and DNA, labelled at one or the other 5′ end with ^{32}P, were digested from the 3′ ends until DNA protected by TIF-1B was encountered. DNA between -21 bp and -7 bp was protected (47). Significantly, a G→A substitution at -16 bp, that reduces promoter activity, reduces the binding of both TIF-1B and TIFD (47),49). Another protein that shares properties with the factor in fraction D is SL1 that has been purified 100,000-fold from Hela cells (51). This protein reprogrammes fractionated mouse cell extracts to transcribe human pre-rRNA genes and so is responsible for promoter selection. However, SL1 is not a sequence-specific DNA binding protein, which suggested that it might be recruited into the preincubation complex though protein-protein interactions with a factor bound to the promoter. This is the case; formation of a stable pre-initiation complex in the human system first requires the sequence-specific binding of another factor (UBF-1) to two sites in the promoter (52). The ability of SL1 to reprogramme the mouse extract was explained by suggesting that the mouse UBF1 equivalent (for which there was evidence) was able to interact with the human SL1.

Extracts from lymphosarcoma cells, treated with glucocorticoids to prevent growth, do not support pre-rRNA biosynthesis unless supplemented with fraction C from growing cells (53). This result implicated a factor in fraction C that controls the rate of pre-rRNA biosynthesis in response to changes in the rate of cell proliferation. Such a factor (TIF-1A) has been partially purified from extracts of growing mouse cells and separated chromatographically from RNA Polymerase I (54). The factor, which cannot be detected in quiescent cells restores transcriptional activity to these extracts.

In similar experiments, Tower *et al.* (55) concluded that transcription of the mouse pre-rRNA gene required two factors C and D. Factor D is identical to TIF-D/TIF-1B whilst factor C was identified as an activated subform of RNA Polymerase I and equated with TIF-1A (56). This subform was abundant in extracts from growing cells and depleted in extracts of down regulated cells. The discrepancy between these results and those of Buttgereit *et al.* (54) was explained as being due to an artifact that arose from the incomplete separation of an inhibitor(s) from RNA polymerase I and that TIF-1A is a mixture of the active subform and this inhibitor(s). The nature of the difference between the two forms of RNA polymerase I is unknown, but is likely to be small as they copurify on various chromatographic resins. The activated subform sediments more quickly than the bulk of RNA polymerase I through sucrose gradients and thus may be larger, or more compact. These findings may be considered in relation to two much earlier observations, namely that rRNA biosynthesis may be controlled by a protein with a rapid turnover (57) and that yeast RNA polymerase I at least is a phosphoprotein the state of which varies with the growth state of the cell (28).

One factor (TIF) and a distinct subform of RNA polymerase I is also sufficient for transcribing the pre-rRNA gene of *Acanthamoeba* (58,59). TIF forms a stable pre-initiation complex with the promoter (46) and in this respect is analogous to TIF-D/TIF-1B. Pre-binding competition experiments using DNA from deletion mutants as the first DNA suggested that TIF bound between -47 bp and -20 bp. Most recent studies suggest that TIF interacts most strongly with DNA centered at -29 bp (60), but a much larger region from -14 bp to -67 bp on the coding strand and from -12 bp to -69 bp on the non-coding strand is protected by TIF from attack by deoxyribonuclease I (59). The TIF footprint is unchanged after transcription is initiated.

Evidence that pre-rRNA biosynthesis in *Acanthamoeba* requires and is controlled by the amount of an activated subform of RNA polymerase I was obtained using extracts and RNA polymerase I from vegetative cells and from encysting cells in which rRNA biosynthesis stops. Extracts of encysting cells do not transcribe the *Acanthamoeba* pre-rRNA gene, but can be programmed to do so by adding RNA polymerase I from vegetative cells (58). However, extracts of encysting cells contain as much RNA polymerase I as vegetative cells when assayed using calf-thymus DNA as a non-specific template. These results, together with observations that the enzyme from encysting cells is more thermolabile, but has the same subunit composition as RNA Polymerase I from vegetative cells, suggest that pre-rRNA biosynthesis during encystment is regulated by subtle modification of the RNA polymerase (58). The inactive form of the enzyme is unable to bind to a TIF-DNA complex (61).

The binding of RNA polymerase I from vegetative cells to a TIF-promoter complex extends the region of DNA protected from deoxyribonuclease I into a second promoter domain around the initiation site (-14 bp to +20bp). This binding occurs independently of the DNA sequence outside the TIF binding site, which suggests that the binding of RNA polymerase I takes place through protein-protein interactions with TIF (62). However, mutations in the region protected by RNA polymerase I prevent transcription, which suggests also that the sequence of DNA here is important for a stage in initiation subsequent to the initial binding of the enzyme. Chemical footprinting and topological analysis showed that RNA polymerase I forms a closed complex with the TIF-promoter pre-initiation complex that is converted into an open complex only when nucleoside triphosphates are added (63).

PROMOTER ELEMENTS

Promoters for pre-rRNA genes show individual characteristics, although each has a proximal element between about -250 bp and +50 bp (64), which may be influenced by enhancing elements from more distal regions of the nontranscribed spacer.

The Proximal Promoter

This may contain two or more domains (64). One domain, the core or minimal promoter, spans the initiation site and is essential for transcription. The other domains modulate the efficiency of expression, but are not active in all transcription assays. Their existence has often been inferred from the results of competition experiments in which the effect of adding wild-type DNA on transcription of mutant rDNA is determined. If the wild-type DNA competes more effectively, transcription of the mutant will be reduced disproportionately.

The pre-rRNA gene of the mouse was the first to be transcribed specifically *in vitro* (24). The promoter consists of three or four domains. In cell extracts in the absence of competitor DNA, 5′ deletions to -40 bp, or to -39bp had no effect on transcription, deletions to -34 bp and -32 bp decreased transcription several fold and deletions to -28 bp and to -27 bp inactivated the gene (65,33,66). Thus, DNA between -39/-40 bp and -34/-32 bp is required for efficient expression and the 5′ boundary of the core promoter sequence is between -34/-32 bp and -27/-28 bp. The 3′ boundary of the core promoter has been placed between +2 and +9 bp (66) and between -1 and +11 bp (33) by analyses of the activities of 3′ deletion mutants. Much of this core domain (~ -40 bp to -1 bp) is required for the binding of TIF-1B/TIF-D, but DNA from -1 bp to +11 bp, although important for transcription, has little effect on the binding of this factor (49). This situation is reminiscent of that in *Acanthamoeba*, which was discussed previously and in which DNA downstream of -14 bp is important for transcription, but is not required for the binding of either TIF, or RNA polymerase I. Bases within the core that may be important for transcription have been analysed using bisulphite mutagenesis. Transitions from C to T at -36, -31, -27, -22 and -21 and -13 had no effect on transcription; a G to A transition at -15 increased transcription 20% and a G to A transition at -16 inactivated the promoter (67). This G residue at -16 bp is required for the binding of the transcription factor TIF-1B to the core promoter sequence (7) and is conserved in several species (Figure 1).

Transcription of the mouse pre-rRNA gene is modulated by a second upstream domain between -39 bp and -169 bp. An effect of this domain was demonstrated in cell extracts only by adding competitor DNA (39), but in transfected mouse cells 5′ deletion mutants with boundaries at -169 bp, or at -320 bp were transcribed ten times more efficiently than a mutant lacking rDNA upstream of -39 bp (68). A region downstream from the initiation site between +11 and +52 bp has also been implicated in modulating the efficiency of transcription (33), although no such effects were detected by Miller *et al.* (66).

A detailed analysis of linker-scanner mutants assayed *in vivo* has shown that the human promoter contains two main elements; a core element between -45 bp and +20 bp that is required for specific transcription and an upstream element (UCE) between -107 bp and -234 bp that may stimulate up to 100-fold (69). This upstream element contains several domains. Extending into the UCE from -81 bp to -110 bp is an element that is important for the binding of UBF-1, which protects DNA between -75 bp and -114 bp from deoxyribonuclease I. Addition of SLl extends the DNA footprint to -165 bp, but because SLl is not a site-specific DNA binding protein it is not clear if the protection is caused by SLl binding, or by a change in UBFl binding (52). A second UBFl binding site is located in the core promoter element.

The activity of domains in the proximal promoter of the pre-rRNA gene of *X. laevis* is influenced both by the nature of the assay and by DNA of the non-transcribed spacer. Two domains were detected by analysing the effects of 5′ and 3′ deletions in recombinants lacking the nontranscribed spacer. DNA from -7 to +6 bp was sufficient for accurate and efficient initiation in oocytes, but in extracts of oocytes supplemented with an ApG primer, DNA from -142 bp to +6 bp was required (70). This upstream domain was required for transcription in oocytes by recombinants that contained an intact upstream nontranscribed spacer (36) and was also required when the concentration of the injected DNA was low (71). Studies of linker-scanner mutants aimed at defining these promoter domains more precisely have yielded conflicting results. According to Windle and Sollner-Webb (72), the upstream domain extends from -140 bp to -128 bp and the core promoter extends from -36 bp to +10 bp. They also found that DNA between -72 bp and -114 bp, which is homologous to DNA found in an enhancer element in the nontranscribed spacer, could be replaced by unrelated DNA without an adverse effect on transcription. By contrast, Reeder *et al.* (73) found that activity was impaired in nearly all of a series of linker scanner mutants that spanned the promoter at approximately 10 bp intervals. Binding sites for proteins have been mapped in the core promoter and in the upstream domain between -70 bp and -100 bp and between -120 bp and -140 bp (74). One protein that protects the core promoter between -5 bp and -30 bp has been identified as nuclear factor I that functions in adenovirus replication and binds to many RNA polymerase II promoters (75). Other pre-rRNA gene promoters also seem to contain NFl binding sites.

The rat promoter conforms to the pattern of a core element (downstream from -31 bp) flanked by an upstream domain including a region between -167 bp and -122 bp that is required for the formation of stable preinitiation complexes (76). The effect of this upstream domain became apparent only in pre-binding competition experiments. Two, or more domains have also been detected in the promoters of the pre-rRNA genes of *D. melanogaster* (77,78) and *A. castellanii*, but in these cases, the additional domains abut the core promoter. Thus in *A. castellanii* DNA between -31 bp and -20 bp is required for TIF binding, DNA between -47 bp and -32 bp is needed to stabilize the TIF-DNA complex and the region between -20 bp and +8 bp is needed for initiation by RNA polymerase I (46,60).

The promoter for pre-rRNA genes in yeast is less well defined, partly because no *in vitro* transcription assay has been developed. For *S. cerevisiae*, analysis of deletion mutants has suggested that DNA within -155 bp to +15 bp is required (79, 115). Three domains within the promoter were identified using linker scanner mutagenesis. Mutations in domains I (+8 bp to -28 bp) and II (-51 bp to -70bp) drastically reduced promoter activity, whilst mutations in domain III (-76 bp to -146 bp) generally had less effect (115). Each of these three domains contains sequences that are conserved in four yeasts (116). The conserved sequence between -9 bp and +23 bp includes and invariant 7 bp sequence from +1 bp to +7 bp inclusive, that is likely to be important for transcription (115). Our data suggest that DNA within -163 bp to -155 bp is also important and that some sequences in domain III between about -120 bp and -110 are essential for activity (M. Butlin, unpublished).

The Role of Enhancers

The *cis*-activation of genes by distant elements of DNA (enhancers) was detected first for genes transcribed by RNA polymerase II. Typically, such enhancers function relatively independently of their distinct, orientation and position upstream or downstream of the promoter. *Trans*-acting proteins bind to sequence motifs within these DNA elements and somehow enhance transcription (80,81,82). Three models have been considered. In two of the models, the proteins are assumed to remain bound at the enhancer. Action at a distance might then result from a topological change that is transmitted to the promoter, or from direct interactions between proteins bound at the enhancer and promoter with the intervening DNA looped out. The third model envisages that the enhancer might act as an entry point for protein factors, which subsequently migrate to the promoter.

Enhancers acting on pre-rRNA genes have been detected in several organisms. To some extent at least, these may act differently from each other and from enhancers that activate genes transcribed by RNA polymerase II.

The enhancing effect of the *Xenopus* nontranscribed spacer was detected initially from the 5 to 10-fold effect of deleting all but 1150 bp of the nontranscribed spacer from copies of the *X. laevis* gene expressed in *X. borealis* embryos (83), and from the competitive advantage enjoyed by recombinants containing the nontranscribed spacer over recombinants lacking part of this DNA, when co-injected into oocytes (84,85). The nontranscribed spacer consists largely of three repetitious regions. In individual rDNA repeats the length of the nontranscribed spacer varies between 2 and 6 kb because the number of copies of the repeated sequences differ. Much of the nontranscribed spacer (repetitive regions 2 and 3, Figure 3) is occupied by direct repeats of 60 and 81 bp elements containing a 28-42 bp sequence that is homologous to DNA in the promoter region between -72 and -114 bp. These 60/81 bp repeats are interrupted every 0.5-1.5 kb by an almost perfect repeat of the promoter from -145 to +4 bp (Figure 3) (86) from which transcription may occur (84). Both the 60/81 bp repeats and the promoter duplication have been implicated in enhancing transcription.

A competitive advantage in co-injection experiments is conferred by nontranscribed spacer DNA containing 60/81 bp repeats, but lacking a spacer promoter. This *cis*-acting effect is independent of orientation, can be transmitted through several kb of plasmid DNA and functions equally well when placed downstream of the promoter (85,87,88). These features resemble those of RNA polymerase II enhancers. Deletion and linker-scanner mutagenesis of the 60/81 bp repeat has shown that a sequence of 56 bp, divided into two

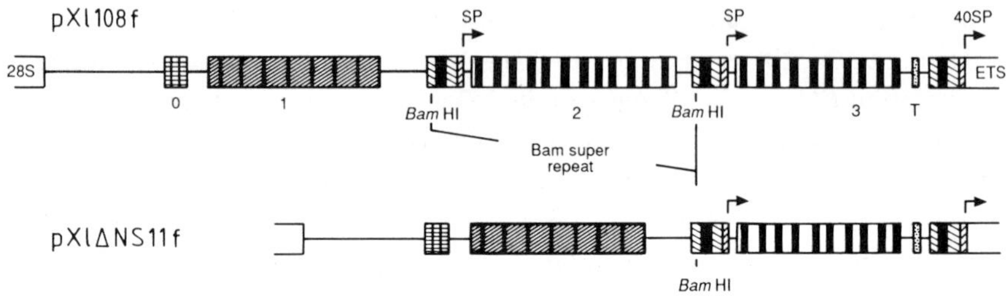

Figure 3: The structure of the nontranscribed spacer of *X. laevis* rDNA in the plasmid pXl108F and of the derivative (pXlΔNS11f) used for the construction of spacer promoter mutants. Reproduced from De Winter and Moss (90). The vertical bars represent copies of the 28-42 b.p. element (see text). 28S, ETS, 40S P and T refer respectively to the 28S rRNA gene, the external transcribed spacer, spacer promoter, 40S pre-rRNA and the termination site. Repetivive regions are numbered 0-3. Reprinted by permission from *Cell* **44**:313-318 Copyright ©1986.

domains, is required for enhancer activity. The larger of these domains corresponds to the upstream promoter element between -72 bp and -114 bp and may bind the same protein (89). The role of the enhancer might be to increase the chance of activating the adjacent promoter, possibly after migration of the bound factor to the promoter.

The role of the spacer promoters was tested by De Winter and Moss (90) using a recombinant (pXLΔNSIIf) with a single spacer promoter and an intact repetitive region 3, but lacking repetitive regions 0, 1 and 2 (Figure 3). Inactivation of the spacer promoter decreased transcription of the pre-rRNA gene approximately 10-fold in *X. borealis* oocytes. This effect was only observed when the spacer promoter was upstream of a set of 60/81 bp repeats, and led to the suggestion that both the spacer promoter and the 60/81 bp repeats are important for enhancement (91). These and other results suggested that the spacer promoter might act by readthrough enhancement in which RNA polymerase I was delivered to the termination site at -213 bp and trapped by the nearby promoter of the next gene. This supply of RNA polymerase would augment delivery from the preceding pre-rRNA gene to the same termination site. In support of the hypothesis of readthrough enhancement, Mitchelson and Moss (92) found that moving the termination site away from the promoter (on a circular plasmid around which RNA polymerase could recycle) greatly reduced the efficiency of transcription. Studies of other organisms have also provided evidence that supports the concept of readthrough enhancement. The nontranscribed spacer of *D. melanogaster* contains repeats of a sequence containing a spacer promoter and enhances transcription of a pre-rRNA minigene *in vitro*. The extent of enhancement correlates with the number of spacer promoters in the enhancer DNA (93). In the rat, an enhancing element was detected between -286 bp and -1018 bp of the nontranscribed spacer that stimulated transcription 17-fold in *in vitro* assays (94). This element contains a spacer promoter for RNA polymerase I that initiates transcription at -713 bp (95) and a binding site for a protein (p16) that is like a high mobility group protein and which stimulates pre-rRNA biosynthesis *in vitro* when the p16 binding site is in *cis* (96). Putative termination sites occur at -167 bp and -635 bp (95). In the mouse too, an RNA polymerase I spacer promoter has been detected at about -2000 bp which transcribes through to a termination site at -171 bp (97,98,99).

The attractive concept of readthrough enhancement received further support from experiments which showed that the terminator just upstream from the promoter augmented transcription from that promoter in mouse and *X. laevis* minigenes on circular plasmids (97,98,100). These results suggested that the terminator might be an essential promoter element and that termination and initiation were concerted. Subsequent experiments on *X. laevis* and mouse minigenes of a similar design to those conducted by Mitchelson and Moss (92) (see above) gave different results. Minigene transcription on a circular plasmid that allows the recycling of RNA polymerase from promoter through to its adjacent terminator was not affected by the introduction of either psoralen adducts (101) or pyrimidine dimers that caused premature termination, nor was transcription affected by separating the terminator from the promoter (102). Thus, in these cases the rate of initiation was unaffected by transcriptional delivery of RNA polymerase to the terminator/promoter complex. Instead, it appears that the role of the terminator is to prevent inhibition of transcription by promoter occlusion (see also 103).

An alternative suggestion is that the *X. laevis* enhancer might be important in the initial formation of an active transcription complex (85,91).

An enhancer for pre-rRNA gene expression in *S cerevisiae* is located in a polymorphic *Eco*RI-*Hind*III fragment of about 190 bp at the 3′ end of the pre-rRNA gene (Figure 2) (43). Type I elements contain a spacer promoter for RNA polymerase I within a 22 bp sequence near the *Hind*III site. In type II elements the spacer promoter has been inactivated by an A→G transition (104,105). Both types of element contain a termination site for pre-rRNA biosynthesis within the polymorphic T-rich region (23). DNA from both extremes of the *Eco*RI-*Hind*III fragment is required for enhancer activity (106,107).

The enhancer in yeast pre-rRNA genes was detected from the reduced rate of transcription in yeast of pre-rRNA minigenes lacking upstream non-transcribed spacer DNA (42,108). Loss of the 190 bp fragment decreased transcription 15-fold in recombinants constructed in YCp50, a centromeric shuttle vector that exists in yeast at about one copy per cell (42). In our experiments, loss of 2.2 kb *Bam*HI fragment, that included the 190 bp enhancer, from a recombinant constructed in the high copy number vector pJDB207 reduced transcription of the minigene 3-fold (108). This difference in the magnitude of the effect of the enhancer on the minigenes constructed in YCp50 and pJDB207 was caused by a second copy of the enhancer at the 3′ end of the minigene in pJDB207.

In their initial experiments, Elion and Warner (42) showed that the 190 bp enhancer was inactive when inverted. More recent studies showed that the dependence on orientation was due to inhibition by neighbouring DNA and that the enhancer may work equally well in either orientation (106). However, some dependence on orientation seems likely as inversion of 2.2 kb and 3.5 kb *Bam*HI fragments containing the enhancer buried deep within, decreased minigene expression significantly (108). The 190 bp enhancer functioned equally well when located downstream of the minigene (106). This positional independence was retained in recombinants integrated into the host genome, which excludes the possibility that the enhancer could be acting as a far upstream enhancer on a circular plasmid molecule.

The coincidence of enhancer, terminator and spacer promoter has inspired several models for enhancer action, each of which invokes a form of readthrough enhancement, but none is proven. In the ribomoter model (Figure 4) the pre-rRNA gene is envisaged as forming a loop joined at the promoter and enhancer/terminator (23). It is proposed that this structure causes RNA polymerase I to recycle. A variant of this model is to propose that DNA of the nontranscribed spacer loops out between the terminator/enhancer and the promoter of the next pre-rRNA gene in this tandemly arrayed set of genes. These models are difficult to reconcile with two observations. The first is that the enhancer works well when situated upstream of a pre-rRNA minigene integrated at the URA 3 locus (106). The second is that the terminator is orientation dependent whereas the enhancer is not (106,107). These observations suggest that delivery of RNA polymerase I to the terminator/enhancer by transcription is not necessary for enhancement, but do not preclude the possibility that the binding of RNA polymerase I to the promoter may be

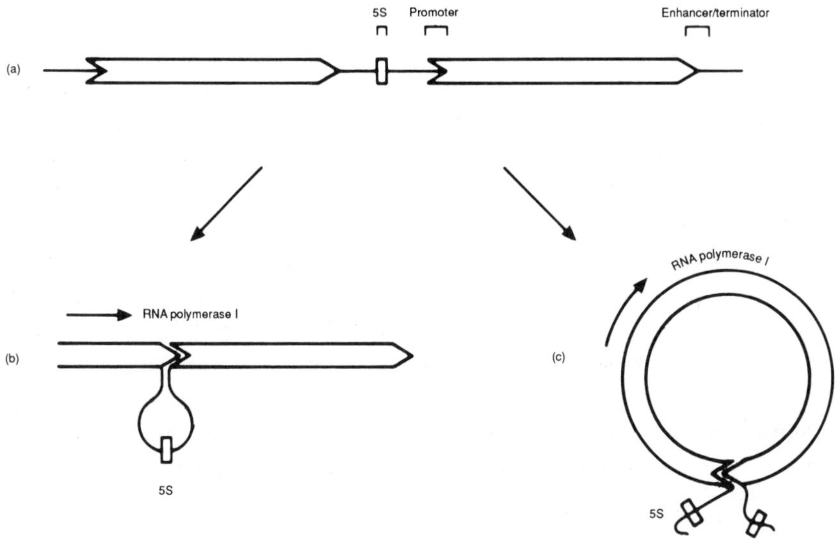

Figure 4: Models for the enhancement of pre-rRNA biosynthesis in yeast. (a) shows 2 tandemly repeated pre-rRNA gened (7 kb. each) separated by nontranscribed spacer DNA (2 kb.) that contains the gene for 5S rRNA (5S). Models for enhancer action envisage that complexes form between the promoter and the enhancer/terminator (23,117). These may form by looping out of the DNA of the nontranscribed spacer (b), or by the pre-rRNA gene itself forming a looping a loop (c). In (b) RNA polymerase I passes from one gene to the next and in (c), the ribomotor model of Kempers-Veenstra *et al.* (1986) (23), the enzyme cycles around a single gene.

facilitated by the formation of an enhancer/promoter complex. These and other observations (106,108), also suggest that the enhancer/terminator does not act by preventing promoter occlusion. A role for the spacer promoter in readthrough enhancement has been suggested, but this suffers from the same drawbacks and also from the fact that both type I and type II elements enhance, but only type I elements have a functional spacer promoter (107). Johnson and Warner (117) have recently proposed that the enhancer/terminator and promoter elements from at least several pre-rRNA genes form a large, loosely organized complex.

CONCLUDING REMARKS

Pre-rRNA genes are unique in being transcribed by an RNA polymerase enzyme dedicated to this task. The rate of transcription is determined by the growth rate of the cell and is highly regulated since the rRNA that is produced is the most abundant RNA in the cell. Always an active field, the study of ribosomal RNA biosynthesis has blossomed in the last few years with the development of *in vivo* and *in vitro* assays for measuring the expression of cloned copies of these genes. Detailed knowledge of the roles of transcription factors in the formation of active transcription complexes is now emerging, as are explanations at the molecular level for the phenomenon of nucleolar dominance (64).

At the cellular level much remains to be learnt of the organisation of pre-rRNA genes in chromatin and their association with the nuclear matrix (109,110,111). The identification of a role for topoisomerase I in the transcription of these genes (112,113) and of binding sites nuclear factor I at their promoters (75) is a further indication of the overall complexity of the process. As these features unfold, the understanding of enhancer action is likely to improve and common mechanistic themes may emerge.

ACKNOWLEDGEMENT

I would like to thank Mike Butlin for many useful suggestions and for much constructive discussion.

REFERENCES

1. Long, E. O, and David, I. B. (1980). *Annu. Rev Biochem.*, **49**:727-764.
2. Labhart, P., and Reeder, R. H. (1986). *Cell*, **45**:431-443.
3. De Winter, R. F. J., and Moss, T. (1986*b*). *Nucleic Acids Res.*, **14**:6041-6051.
4. Tautz, D., and Dover, G. A. (1986). *EMBO J.*, **5**:1267-1273.
5. Beckingham, K. (1982) **In:** *The Cell Nucleus : rDNA Part A* (Busch, H., and Rothblum, L. Eds.) Academic Press, New York, pp. 206-269.
6. Cech, T. R., Zang, A. J., Grabowski, P. J., and Brehm, S. L. (1982) **In**: *The Cell Nucleus : rDNA Part A,* (Busch, H., and Rothblum, L., Eds.) Academic Press, New York, pp. 171-204.
7. Cech, T. R. (1986). *Cell*, 44:207-210.
8. Perry, R. P. (1976). *Annu. Rev. Biochem.*, **45**:605-630.
9. Berk, A. J., and Sharp, P. A. (1977). *Cell*, **12**:721-732.
10. Casey, J., and Davidson, N. (1977). *Nucleic Acids Res.*, **4**:1539-1552.
11. Reeder, R. H., Sollner-Webb, B., and Wahn, H. L. (1977). *Proc. Natl. Acad. Sci USA*, **74**:5402-5406.
12. Financsek, I., Mizumoto, K., and Muramatsu, M. (1982) *Gene*, **18**:115-122.
13. Financsek, I., Mizumoto, K., Michima, Y., and Muramatsu, M. (1982) *Proc. Natl. Acad. Sci USA*, **79**:3092-3096.
14. Nikolaev, N., Georgiev, O. J., Venkov, P. V., and Hadjiolov, A. A. (1979). *J. Mol. Biol.*, **127**:297-308.
15. Klootwijk, J., de Jonge, P., and Planta, R. J. (1979) *Nucleic Acids Res.*, **6**:27-39.

16. Levis, R., and Penman, S. (1978). *J. Mol. Biol.*, **121**:219-238.
17. Bach, R., Grummt, I, and Allet, B. (1981). *Nucleic Acids Res.*, **9**:1559-1569.
18. Batts-Young, B., and Lodish, H. (1978). *Proc. Natl. Acad. Sci. USA*, **75**:740-744.
19. Kuhn, A., Normann, A., Bartsch, I., and Grummt, I. (1988). *EMBO J.*, **7**:1497-1502.
20. Kermekchiev, M. B., and Grummt, I. (1987). *Nucleic Acids Res.*, **15**:4131-4143.
21. Kuhn, A., and Grummt, I. (1989). *Genes and Development*, **3**:224-231.
22. Labhart, P., and Reeder, R. H. (1987). *Cell*, **50**:51-57.
23. Kempers-Veenstra, A. E., Oliemans, J., Offenberg, H., Dekker, A. F., Piper, P. W. Planta, R. J., and Klootwijk, J. (1986). *EMBO J.*, **5**:2703-2710.
24. Grummt, I. (1981). *Proc. Natl. Acad. Sci. USA*, **78**:727-731.
25. Bakken, A., Morgan, G., Sollner-Webb, B., Roan, J., Busby, S., and Reeder, R. H. (1982). *Proc. Natl. Acad. Sci. USA*, **79**:56-60.
26. Kohorn, B. D., and Rae, P. M. (1982). *Proc. Natl. Acad. Sci. USA*, **79**:1501-1505.
27. Miesfeld, R., and Arnheim, N. (1984). *Mol. Cell. Biol.*, **4**:221-227.
28. Sentenac, A. (1985). **In:** *Critical Reviews in Biochemistry*, **18**:31-90.
29. Beebee, T. J. C., and Butterworth, P. H. W. (1974). *Eur. J. Biochem.*, **45**:395-406.
30. Van Keulen, H., and Retel, J. (1977). *Eur. J. Biochem.*, **79**:579-588.
31. Weil, P. A., Segall, J., Harris, B., Ng, S. Y., and Roeder, R. G. (1979). *J. Biol. Chem.*, **254**:6163-6173.
32. Manley, J. L., Fire, A., Cano, A., Sharp, P. A., and Gefter, M. L. (1980). *Proc. Natl. Acad. Sci. USA*, **77**:3855-3859.
33. Yamamoto, O., Takakusa, N. Michima, Y., Kominami, R., and Muramatsu, M. (1984). *Proc. Natl. Acad. Sci. USA*, **81**:299-303.
34. Trendelenburg, M., and Gurdon (1978). *Nature*, **276**:292-294.
35. Sollner-Webb, B., and McKnight, S. L. (1982). *Nucleic Acids. Res.*, **10**:3391-3405.
36. Moss, T. (1982). *Cell*, **30**:835-842.
37. Vance, V. B., Thompson, E. A., and Bowman, L. H. (1985). *Nucleic Acids Res.*, **13**:7499-7513.
38. Smale, S. T., and Tjian, R. (1985). *Mol. Cell. Biol.*, **5**:352-362.
39. Grummt, I., Roth, E., and Paul, M. R. (1982). *Nature* (London), **296**:173-174.
40. Quincey, R. V., and Arnold, R. E. (1984). *Biochem. J.*, **224**:497-503.
41. Kempers-Veenstra, A. E., van Heerikhuizen, H., Muster, W., Klootwijk, S., and Planta, R. J. (1984). *EMBO J.*, **3**:1377-1382.
42. Elion, E. A., and Warner, J. R. (1984). *Cell*, **39**:663-673.
43. Wilkinson, J. K., and Sollner-Webb, B. (1982). *J. Biol. Chem.*, **257**:14375-14383.
44. Wandelt, C., and Grummt, I. (1983). *Nucleic Acids Res.*, **11**:3795-3809.
45. Cizewski, V., and Sollner-Webb, B. (1983). *Nucleic Acids. Res.*, **11**:7043-7056.
46. Iida, C. T., Kownin, P., and Paule, M. R. (1985). *Proc. Natl. Acad. Sci. USA*, **81**:1668-1672.
47. Clos, J., Buttgereit, D., and Grummt, I. (1986). *Proc. Natl. Acad. Sci. USA*, **83**:604-608.
48. Mishima, Y., Financsek, L., Kominami, R., and Muramatsu, M. (1982). *Nucleic Acids Res.*, **10**:6659-6670.
49. Nagamine, M., Kishimoto, T., Aono, J., Kato, H., Kominami, R., and Muramatsu, M., (1987). *Mol. Cell. Biol.*, **7**:1486-1495.
50. Safrany, G., Tanaka, N., Kishimoto, T., Ishikawa, Y., Kato, H., Kominami, R., and Muramatsu, M. (1989). *Mol. Cell Biol.*, **9**:349-353.
51. Learned, R. M., Cordes, S., and Tjian, R. (1985). *Mol. Cell. Biol.*, **5**:1358-1369.
52. Bell, S. P., Learned, M., Jantzen, H-M., and Tjian, R. (1988). *Science*, **241**:1192-1197.
53. Cavanagh, A. H., and Thompson, A. E. (1985). *Nucleic Acids Res.*, **13**:3357-3369.
54. Buttgereit, D., Pflugfelder, G., and Grummt, I. (1985). *Nucleic Acids Res.*, **13**:8165-8180.
55. Tower, J., Culotta, V. C., and Sollner-Webb, B. (1986). *Mol. Cell. Biol.*, **6**:3451-3462.
56. Tower, J., and Sollner-Webb, B. (1987). *Cell*, **50**:873-883.
57. Yu, F. L., and Fiegelson, P. (1971). *Proc. Natl. Acad. Sci USA*, **68**:2177-2180.

58. Paule, M. R. Iida, C. T., Perna, P. J., Harris, G. H. Knoll, D. A., and Alessio, J. M. (1984). *Nucleic Acids Res.*, **12**:8161-8180.
59. Bateman, E., Iida, C. T., Kownin, P., and Paule, M. R. (1985). *Proc. Natl. Acad. Sci. USA*, **82**:8004-8008.
60. Kownin, P., Bateman, E., and Paule, M. R. (1988). *Mol. Cell. Biol.*, **8**:743-753.
61. Bateman, E., and Paule, M. R. (1986). *Cell*, **47**:445-450.
62. Kownin, P., Bateman, E., and Paule, M. R. (1987). *Cell*, **50**:693-699.
63. Bateman, E., and Paule, M. R. (1988). *Mol. Cell Biol.*, **8**:1940-1946.
64. Sollner-Webb, B., and Tower, J. (1986). *Annu. Rev. Biochem.*, **55**:801-830.
65. Grummt, I. (1982). *Proc. Natl. Acad. Sci. USA*, **79**:6908-6911.
66. Miller, K. G., Tower, J., and Sollner-Webb, B. (1985). *Mol. Cell Biol.*, **5**:554-562.
67. Skinner, J. A., Ohrlein, A., and Grummt, I. (1984). *Proc. Natl. Acad. Sci. USA*, **81**:2137-2141.
68. Grummt, I., and Skinner, J. A. (1985). *Proc. Natl. Acad. Sci. USA*, **82**:722-726.
69. Jones, M. H., Learned, R. M., Tijan, R. (1988). *Proc. Natl. Acad. Sci. USA*, **85**:669-673.
70. Sollner-Webb, B. Wilkinson, J. A. K., Roan, J., and Reeder, R. H. (1983). *Cell*, **35**:199-206.
71. Windle, J. L., and Sollner-Webb, B. (1986). *Mol. Cell Biol.*, **6**:1228-1234.
72. Windle, J. L., and Sollner-Webb, B. (1986). *Mol. Cell Biol.*, **6**:4585-4593.
73. Reeder, R. H., Pennock, D., McStay, B., Roan, J., Tolentino, E., and Walker, P. (1987). *Nucleic Acids Res.*, **15**:7429-7441.
74. Dunaway, M., and Reeder, R. H. (1985). *Mol. Cell Biol.*, **5**:313-319.
75. Walker, P., and Reeder, R. M. (1988). *Nucleic Acids Res.*, **16**:10657-10668.
76. Cassidy, B., Haglund, R., and Rothblum, L. I. (1987). *Biochem. Biophys. Acta.*, **909**:133-144.
77. Kohorn, B. D., and Rae, P. M. M. (1983). *Proc. Natl. Acad. Sci. USA*, **80**:3265-3268.
78. Kohorn, B. D., and Rae, P. M. M. (1983). *Nature* (London), **304**:179-181.
79. Kempers-Veenstra, A. E., Musters, W., Dekker, A. F., Klootwijk. J., and Planta, R. J. (1985). *Curr. Genetics*, **10**:253-260.
80. Wasylyk, B. (1988). *Biochem. Biophys. Acta.*, **951**:17-35.
81. Jeang, K-T., and Khoury, G. (1988). *Bioassays*, **8**:104-107.
82. Ptashne, M. (1988). *Nature*, **335**:683-689.
83. Busby, S. J., and Reeder, R. H. (1983). *Cell*, **34**:989-996.
84. Moss, T. (1983). *Nature* (London), **302**:223-228.
85. Reeder, R. H., Roan, R. J., and Dunaway, M. (1983). *Cell*, 35:449-456.
86. Boseley, P., Moss, T., Machler, M., Portman, R., and Birnstiel, M. L. (1979). *Cell*, **17**:19-31.
87. Labhart, P., and Reeder, R. H. (1984). *Cell*, **37**:285-289.
88. Labhart, P., and Reeder, R. H. (1985). *Nucleic Acids Res.*, **13**:8999-9009.
89. Pickaard, C. S., and Reeder, R. H. (1988). *Mol. Cell Biol.*, **8**:4282-4288.
90. De Winter, R. F. J., and Moss, T. (1986a). *Cell*, **44**:313-318.
91. De Winter, R. F. J., and Moss, T. (1987). *J. Mol. Biol.*, **196**:813-827.
92. Mitchelson, K., and Moss, T. (1987). *Nucleic Acids Res.*, **15**:9577-9596.
93. Grimaldi, G., and Di Nocera, P. P. (1988). *Proc. Natl. Acad. Sci. USA*, **85**:5502-5506.
94. Cassidy, B. G., Yan-Yen, H-F., and Rothblum, L. I. (1986). *Mol. Cell Biol.*, **6**:2766-2733.
95. Cassidy, B. G., Yang-Yeu, H-F., and Rothblum, L. I. (1987b). *Mol. Cell Biol.*, **7**:2388-2396.
96. Yang-Yen, H-F., and Rothblum, L. I. (1988). *Mol. Cell Biol.*, **8**:3406-3414.
97. Grummt, I., Kuhn, A., Bartsch, I., and Rosenbauer, H. (1986). *Cell*, **47**:901-911.
98. Henderson, S., and Sollner-Webb, B. (1986). *Cell*, **47**:891-900.
99. Kuhn, A., and Grummt, I. (1987). *EMBO J.*, **6**:3487-3492.
100. McStay, B., and Reeder, R. H. (1986). *Cell*, **47**:913-920.
101. Lucchini, R., and Reeder, R. H. (1989). *Nucleic Acids Res.*, **17**:373-387.
102. Henderson, S. L., Ryan, K., and Sollner-Webb, B. (1989). *Genes and Development*, **3**:212-223.
103. Bateman, E., and Paule, M. R. (1988). *Cell*, **54**:985-992.
104. Swanson, M. E., and Holland, M. J. (1983). *J. Biol. Chem.*, **258**:3242-3250.

105. Swanson, M. E., and Holland, M. J. (1985). *J. Biol. Chem.*, **260**:9905-9915.
106. Elion, E. A., and Warner, J. R. (1986). *Mol. Cell Biol.*, **6**:2089-2097.
107. Mestel, R. Yip, M., Holland, J. P., Wang, E., Kang, J., and Holland, M. J. (1989). *Mol. Cell Biol.*, **9**:1243-1254.
108. Quincey, R. V., and Godfrey, R. E. (1985). *Biochem. J.*, **232**:205-209.
109. Udvardy, A., Louis, C., Han, S., and Schedl, P. (1984). *J. Mol. Biol.*, **175**:113-130.
110. Ness, P. J., Parish, R. W., and Keller, T. (1986). *J. Mol. Biol.*, **188**:287-300.
111. Keppel, F. (1986). *J. Mol. Biol.*, **187**:15-21.
112. Garg, L. C., DiAngelo, S., and Jacob, S. T. (1987). *Proc. Natl. Acad. Sci. USA*, **84**:3185-3188.
113. Zhang, H., Wang, J. C., and Liu, L. F. (1988). *Proc. Natl. Acad. Sci. USA*, **85**:1060-1064.
114. Sommerville, J. (1984). *Nature* (London), **310**:189-190.
115. Musters, W., Knot, J., Maas, P., Dekker, A. F., van Heerikhuizen, H. and Planta, R. J. (1989) *Nucleic Acids Res.* **17**: 9661-9667.
116. Verbeet, M. P., Klootwijk, J., van Heerikhuizen, H., Fontijn, R. D., Vreugdenhil, E., and Planta, R. J. (1984). *Nucleic Acids Res.*, **12**:1137-1148.
117. Johnson, S. P. and Warner, J. R. (1989) *Mol. Cell. Biol.*, **9**: 4986-4993.

6

A REGULATORY MECHANISM IN MAMMALIAN CELLS RESPONDS TO DEACYLATION OF tRNA AND RIBOSOMAL IDLING[1]

Oscar A. Scornik

Department of Biochemistry
Dartmouth Medical School
Hanover, NH 03756

1 I welcome the opportunity to express in this volume my lasting gratitude to Mahlon B. Hoagland. I owe my interest in the regulation of protein metabolism to my association with him, more than two decades ago. The work in liver regeneration which I had started with Mahlon led to our finding of the previously unsuspected role of protein degradation in the control of cellular protein content. The research described here explores a new regulatory function of tRNAs, a family of molecules discovered by Mahlon and his colleagues in 1957.

Unicellular organisms are exposed to changes in the environment in which they live. They compensate for the scarcity of some substances by their more efficient transport or increased biosynthesis and adjust their growth rate to the availability of nutrients. In higher organisms, tissue specialization and hormonal regulation maintain relatively constant concentrations of nutrients in the body fluids. Individual cells in these organisms require fewer direct regulatory responses to the concentration of nutrients; notable among them is the response to amino acid deprivation. When the concentration of a required amino acid in the medium is suboptimal, these cells exhibit a variety of metabolic changes, including slower synthesis of RNA and protein, faster protein breakdown and arrest of growth. The response is very similar to that in bacteria, where the physiological changes following deprivation of nitrogen or a required amino acid are collectively known as the "stringent response".

The bacterial stringent response is the result of a regulatory mechanism sensitive to the level of amino acylation of tRNAs. This article reviews the evidence for the existence of a similar mechanism in cultured CHO cells.

BACTERIAL STRINGENT RESPONSE

The response of *Escherichia coli* to amino acid deprivation is much better understood than in mammalian cells and provides a useful frame of reference for experimentation. The stringent response comprises a number of metabolic changes, including faster protein degradation and slower synthesis of protein and RNA, elicited by deprivation of a required amino acid. This response became understood over the last two decades by investigators interested primarily in control of RNA synthesis (see 8,9,25). More recently it has been extended to the control of protein breakdown (88). The regulatory mechanism recognizes the level of aminoacylation of tRNA. The response is triggered by

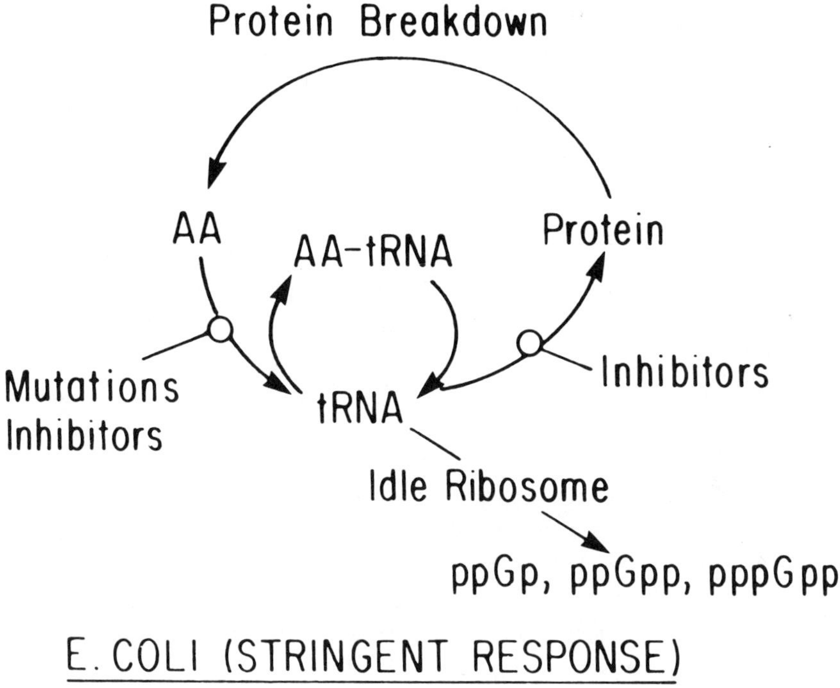

Figure 1: Schematic view of the bacterial stringent response.

the interaction of deacylated tRNA with idle ribosomes, *i.e.*, ribosomes with a free aminoacyl site which do not elongate their nascent chains for lack of the corresponding aminoacyl-tRNA (33). When the deacylated species binds instead, a regulatory signal is generated. A protein, the stringent factor, is necessary for this process. The signal is guanosine 5′-diphosphate, 3′-diphosphate (ppGpp), a member of a family of related nucleotides, including guanosine 5′-diphosphate, 3′-phosphate (ppGp) and guanosine 5′-triphosphate, 3′-diphosphate (pppGpp) (8,58) (Figure 1).

The major lines of evidence for this scheme are as follows.

(a) The effect of amino acid starvation is mimicked by inhibitors or mutations of the aminoacyl-tRNA synthetases, and counteracted by inhibitors of protein synthesis that reduce the utilization of aminoacyl-tRNA (53). These observations indicate that the regulation is triggered by changes in the levels of aminoacyl-tRNA rather than the intracellular concentration of the amino acids or the rate of protein synthesis *per se*.

(b) The response coincides with sharp increases in the cellular concentration of ppGpp. Mutants lacking a functional stringent factor (rel A-), in which the appearance of ppGpp is blocked, fail to exhibit a decreased rate of rRNA synthesis in the absence of a required amino acid (5,87). In other mutants (spoT$^-$), when the amino acid is restored ppGpp disappears only slowly; in these mutants rRNA and protein synthesis take correspondingly longer to recover (86,46). The correlation between the stringent response and the appearance of ppGpp can be demonstrated in a variety of experimental conditions (8,25) but some exceptions have been reported (25).

(c) The synthesis of ppGpp can be reproduced in cell-free systems. It requires the non-enzymatic binding of a deacylated tRNA to the aminoacyl site of ribosomes idling in the corresponding codon, GTP, ATP and stringent factor (the product of the rel A gene) (33).

(d) ppGpp inhibits the transcription of the isolated rRNA gene *in vitro*. The inhibition is observed with only one of two possible promoters of this gene (69,41). Effects of ppGpp on the *in vitro* transcription of other genes have also been described (see 25). There is at present no indication of whether ppGpp can affect directly biochemical processes other than transcription.

EFFECTS OF AMINO ACID DEPRIVATION IN MAMMALIAN CELLS

When amino acid starvation is produced in perfused organs, isolated or cultured cells, a number of changes can be observed. Some may be a direct consequence of the lack of amino acids as substrates, for instance a decreased rate of elongation of ribosomal nascent chains. Other changes reflect regulatory adjustments of cellular growth, for example a decreased rate of RNA synthesis (22,37,19,31) or an inhibition of translational initiation leading to disaggregation of polyribosomes (38,65,91,92,85,13). Still others represent attempts by the cell to compensate for the starvation, notably an accelerated protein degradation (37,98,7,30,93,40,78,76) which, by providing an endogenous source of amino acids, permits continuing synthesis of the most vital proteins and metabolites. The regulatory response is strikingly similar to the bacterial stringent response but ppGpp has not been found in mammalian cells (see 82).

Many attempts to demonstrate the regulatory role of tRNA have been negative or subject to criticism. Inhibition of histidyl-tRNA synthetase by histidinol, a potent competitive inhibitor (92), has been reported to decrease rates of rRNA synthesis in cultured mammalian cells (31,95). No attempt was made to compare the effects of histidinol with those of other inhibitors of protein synthesis. It is thus not clear whether this decrease in rRNA synthesis was related specifically to the deacylation of histidyl-tRNA or, more generally, to the inhibition of protein synthesis, which is known also to result in slower synthesis of rRNA (23). The distinction is even more difficult to make in studies of polyribosome disaggregation (92,95) where inhibitors of protein synthesis other than histidinol directly affect the very parameters one wishes to measure.

Effects on protein degradation are easier to interpret. Whereas amino acid starvation stimulates degradation, inhibitors of protein synthesis have, if anything, the opposite effect. The response of amino acid withdrawal cannot be thus confused with the slower protein synthesis. Nevertheless, neither treatment with histidinol (4) nor inactivation of a synthetase by temperature-sensitive mutations (85,12,32) could be found to stimulate protein degradation in Chinese hamster ovary (CHO) cells. These results, and a similar lack of effect of histidinol on protein degradation in mouse hepatocytes (77), suggested that in mammalian cells the increase in protein breakdown may not depend on the amino acylation of tRNA. These and other experiments (2,21,90) have raised the possibility that there is no relationship between the utilization of amino acids for protein synthesis and the regulation of protein metabolism in mammalian cells. As a result, attention has turned in other directions, such as regulation by amino acid catabolism (90), multiple recognition mechanisms of some amino acids and their metabolites (64,63) or the generation of ammonia (78).

RECOGNITION OF THE LEVEL OF AMINO ACYLATION OF $tRNA^{His}$ IN MAMMALIAN CELLS

Work from our laboratory has provided evidence that histidine starvation of cultured CHO cells activates a regulatory mechanism that recognizes the steady-state level of aminoacylation of $tRNA^{His}$ (76,70)(Figure 2). We selected this model because there are CHO cell lines with temperature-sensitive mutations in histidyl-rRNA synthetase, and we can also inhibit the enzyme with a powerful competitive inhibitor, histidinol (92). Histidine is not synthesized in mammalian cells and serves no obvious metabolic function other than as a precursor for protein synthesis. A further advantage is that the genetic code includes only one wobble pair of codons for histidine (CAU and CAC), which minimizes the complication of isoacceptor tRNAs.

Either the mutants or histidinol can mimic the effect of histidine starvation. These effects are counteracted by cycloheximide which permits the accumulation of the aminoacylated species. Our basic measurement is the

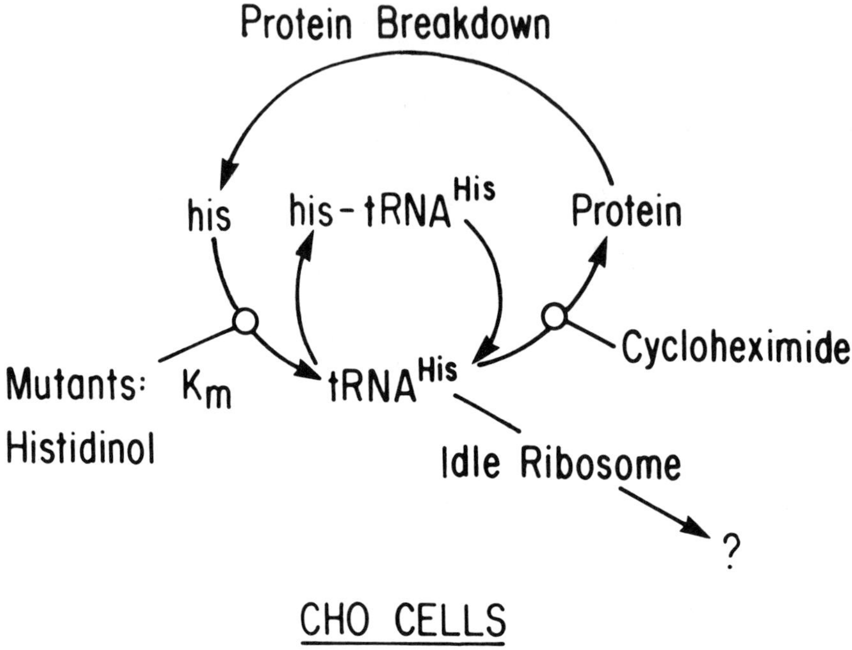

Figure 2: Schematic view of the mammalian regulatory mechanism summarized in this review.

increase in protein degradation elicited in these cells by amino acid starvation. Removal of all amino acids stimulates protein degradation by about 70-80%. These effects are independent of, and additive to, the stimulation obtained by removal of serum (76). We prefer to study the cells in the absence of serum because removal of either serum or amino acids results in growth arrest. By studying amino acid starvation in the absence of serum we ensure that the cells are always quiescent, and we can observe the effects of amino acids uncomplicated by changes in growth conditions. Removal of just one amino acid, such as histidine, produces a smaller but still measurable stimulation of about 20%. Mutations of histidyl-tRNA synthetase mimic this effect. The obvious way to use these temperature-sensitive mutants is to warm them up to the nonpermissive temperature. But elevated temperature has an effect of its own on degradation, which has obscured the interpretation of previous attempts (85,12,32). We have taken a different approach. We knew that although enzymes carrying a temperature-sensitive mutation are active at the low, permissive temperature, they often exhibit abnormal properties, such as

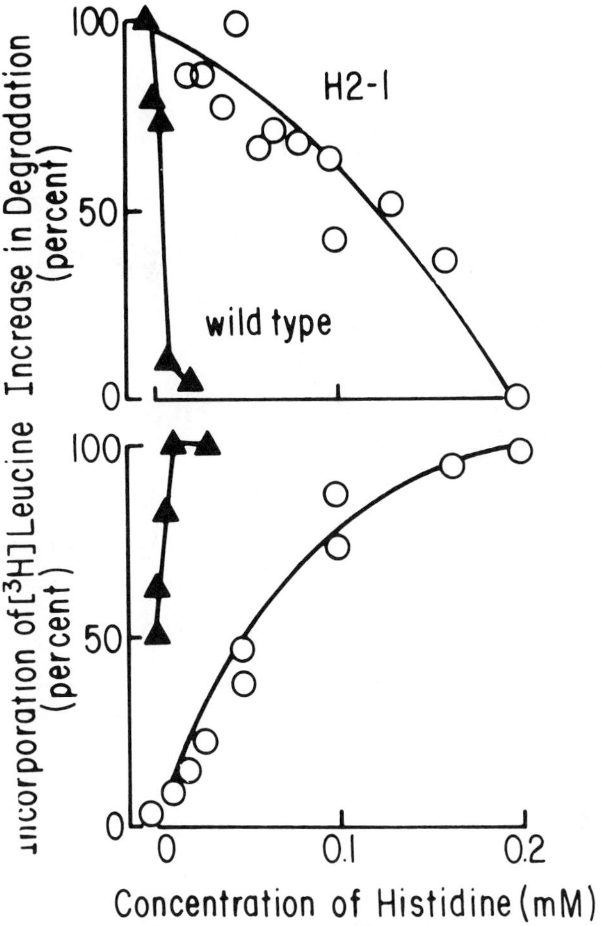

Figure 3: Effects of the concentration of histidine (*abscissa*) on protein synthesis (*bottom ordinate*) and degradation (*top ordinate*) in wild-type CHO cells (*filled triangles*) and cell line H2-1 (*open circles*) carrying a mutation in the histidyl-tRNA synthetase characterized by higher K_m for histidine (adapted from 76).

higher K_m for the substrates. We found two cell lines that required higher than normal concentrations of histidine for optimum protein synthesis because of a higher K_m for histidine of the altered enzyme. We reasoned that if the regulatory mechanism recognizes the amino acylation of tRNA, the cells must also need higher concentrations of histidine for full repression of protein

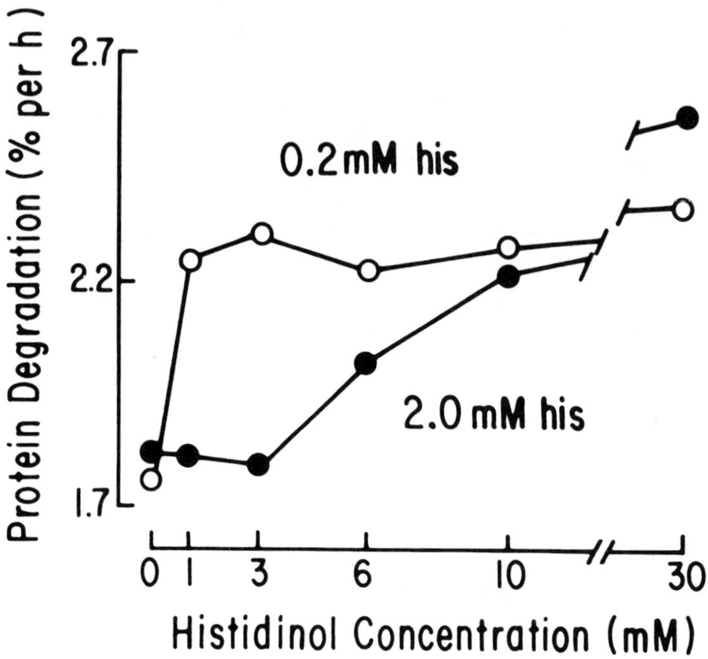

Figure 4: Effects of different concentrations of histidinol (*abscissa*) on protein degradation (*ordinate*) in CHO cells, in the presence of 0.2 mM (*open circles*) and 2.0 mM (*filled circles*) histidine. (From 70).

breakdown. This is what we found in both cell lines; one of them is shown in Figure 3 (76).

Another way to interfere with the synthesis of histidyl-tRNA is with histidinol. Histidinol mimics the effect of histidine starvation. Stimulation of protein breakdown by histidinol results from competition with histidine, because if the concentration of histidine is increased 10-fold, a comparatively larger concentration of histidinol is required (Figure 4).

These effects are not caused by slower protein synthesis. Inhibition of protein synthesis does not stimulate degradation. On the contrary, in this as in many other cells, cycloheximide counteracts the effects of histidine starvation. The stimulatory effects of histidinol and mutations show that regulation depends on the aminoacylation of tRNA, or some consequence of it other than protein synthesis itself. They do not prove, however, that the level of aminoacylation is the important variable. It could be that some activity of the synthetase itself triggers the reaction (61). For instance, these enzymes are

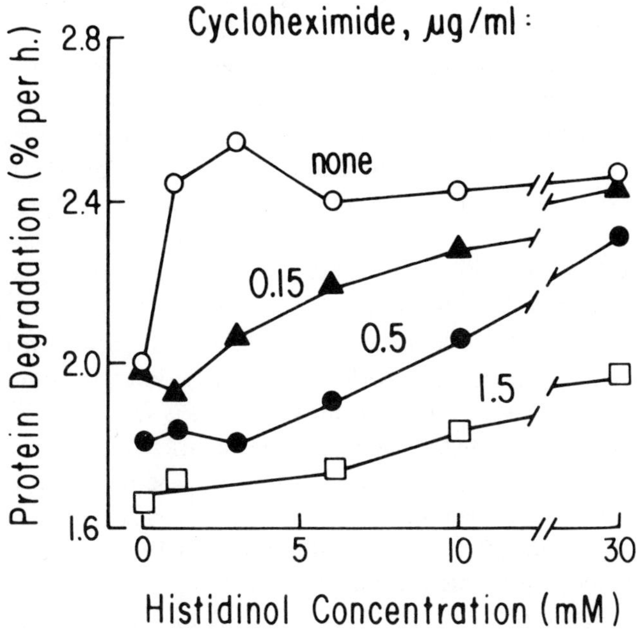

Figure 5: Effects of different histidinol concentrations (*abscissa*) on protein degradation (*ordinate*) in CHO cells in the absence (*open circles*) or presence of 0.15 (*filled triangles*), 0.5 (*filled circles*) or 1.5 (*open squares*) µg/ml of cycloheximide. Concentration of histidine in the medium was 0.2 mM. (From 70).

capable of synthesizing a powerful regulatory dinucleotide, diadenosine 5′5‴-P^1,P^4-tetraphosphate (Ap4A) (66). The following experiments indicate that the steady-state level of aminoacylation of $tRNA^{His}$ is the critical variable in this process.

We ask whether cycloheximide counteracts the effects of histidine starvation and histidinol because it slows down the utilization of histidyl-tRNA for protein synthesis, and thus permits its accumulation or prevents the buildup of the deacylated species. If the hypothesis is correct, we should be able to find a range of critical concentrations of histidinol and cycloheximide where the effects of one are counteracted by a larger concentration of the other. We first define such a range of concentrations for both inhibitors by directly measuring their combined effect on steady-state levels of histidyl-tRNA. In this range we find that the effects of the inhibitors on protein degradation also counteract each other (Figure 5), *i.e.*, the effect of low concentrations of histidinol is reversed by cycloheximide but can be reestablished with higher concentrations of histidinol (70).

In light of the preceding discussion, it seems clear that the regulatory mechanism recognizes neither the rate of aminoacylation of tRNA nor its rate of utilization for protein synthesis, but the steady-state level of the intermediates: histidyl-tRNA and/or the deacylated species (Figure2). In this the response of CHO cells to histidine starvation parallels the stringent response in *E. coli*.

Some effects of amino acids on cellular protein metabolism may be unrelated to their utilization for protein synthesis (2,21,64,90). In some experiments the effects occur in the absence of measurable changes in the levels of the corresponding aminoacyl-tRNAs (see 21). Interpretation of these results should be cautious because if the regulation is successful the changes may be subtle, and because the pools of amino acyl-tRNAs are very small and turn over very fast, on the order of 1-2 s (56,70). The measured value may depend on how rapidly this process is stopped. In attached or suspended cells, the preferred procedure is to add acid to them (89,56). With organs, even close contact with metal precooled in liquid N_2 (21) may not be fast enough. If deacylation of one or more tRNAs is involved, the effects of amino acid starvation should be reversed by inhibitors of protein synthesis, such as cycloheximide, which minimize the utilization of aminoacyl-tRNAs and therefore permit the accumulation of the fully aminoacylated species. This is what happens in the experiments described above. Stimulation of protein degradation by amino acid deprivation in perfused livers is also reversed by cycloheximide (43,51). One can also observe, however, effects of amino acids in the presence of cycloheximide, for instance inhibition of muscle protein breakdown by branched-chain amino acids (90) or inhibition by lysine starvation of translational initiation in Erlich ascites tumor cells (2). In CHO cells, deprivation of individual essential amino acids stimulates protein breakdown only in the absence of cycloheximide, but the larger stimulation caused by removing all amino acids from the medium is partially insensitive to cycloheximide (76).

Cycloheximide-insensitive effects of amino acids may be linked to their catabolism. For example, if hepatocytes are prevented from synthesizing urea, ammonia accumulates, which inhibits protein degradation (78). This is probably not the physiological reason why amino acids suppress degradation; we have shown effects of amino acids in hepatocytes in the absence of ammonia accumulation (unpublished observations). In other instances, the effects of some amino acids could be due to their role as an energy source or as precursors for other processes such as gluconeogenesis (63). The

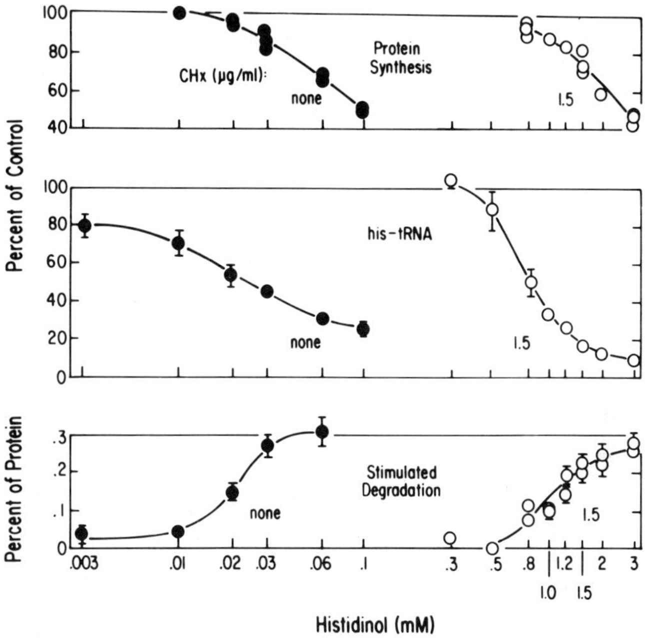

Figure 6: Effects of histidinol on protein synthesis (*top*), levels of his-tRNA (*middle*) (both as % of control) and on protein degradation (*bottom*) (the latter as stimulation above control in % of protein per h), in the absence (*filled circles*) or presence (*open circles*) of cycloheximide, 1.5 µg/ml. Concentration of histidine in the medium was 0.02 mM. (From 74).

physiological role of these effects in the whole animal have yet to be defined (see G).

ROLE OF IDLE RIBOSOMES IN CHO CELLS

As explained in section B, in *E. coli* ppGpp results from the interaction of deacylated tRNA with idle ribosomes, *i.e.* those with a free amino acyl site which cannot elongate their nascent chains for lack of an amino acyl-tRNA . We now have evidence that idle ribosomes are also necessary for the regulation in CHO cells (74). The evidence is based on the quantitative comparison of the effects of histidinol on protein synthesis, degradation and levels of his-tRNA in the absence or presence of cycloheximide, 1.5 µg/ml. At this concentration, cycloheximide inhibits protein synthesis (and hence the rate of utilization of his-tRNA) to 6% of control. Deacylation of $tRNA^{His}$ and stimulation of protein breakdown still can be obtained if sufficiently high

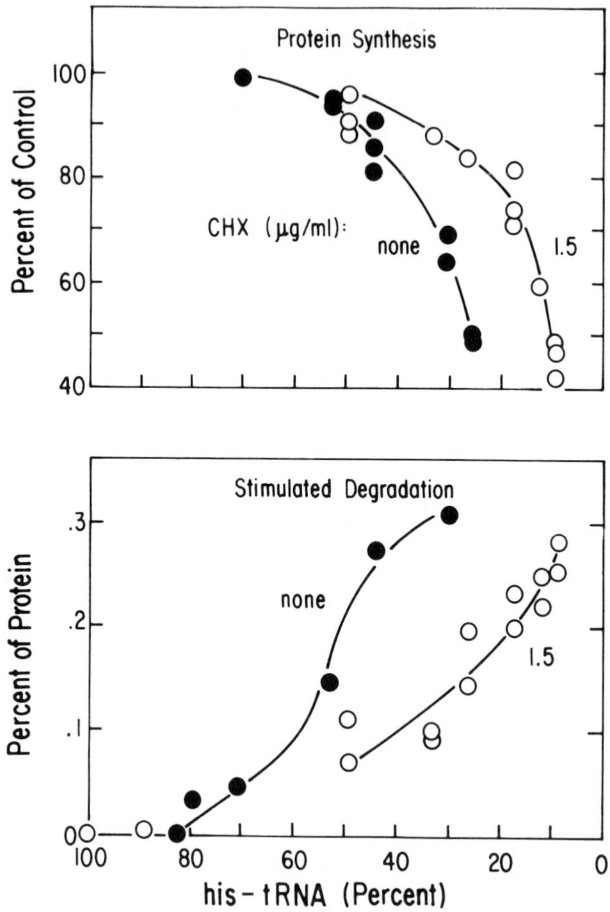

Figure 7: Effects of depletion of his-tRNA on protein synthesis (*top*) and stimulated degradation (*bottom*) in the absence (*filled circles*) and presence (*open circles*) of cycloheximide, 1.5 μg/ml. Same values as in Figure 6, replotted. (From 74).

histidinol/histidine ratios are used. These high ratios can be attained with reasonable levels of histidinol by lowering ten-fold the concentration of histidine in the medium (Figure 6).

By plotting the effects of histidinol on protein synthesis and degradation against the levels of his-tRNA (Figure 7) we now focus on two observations: (a) Cycloheximide dissociates the effects of histidinol on his-tRNA from its effects on protein synthesis. In the presence of cycloheximide, a more exten-

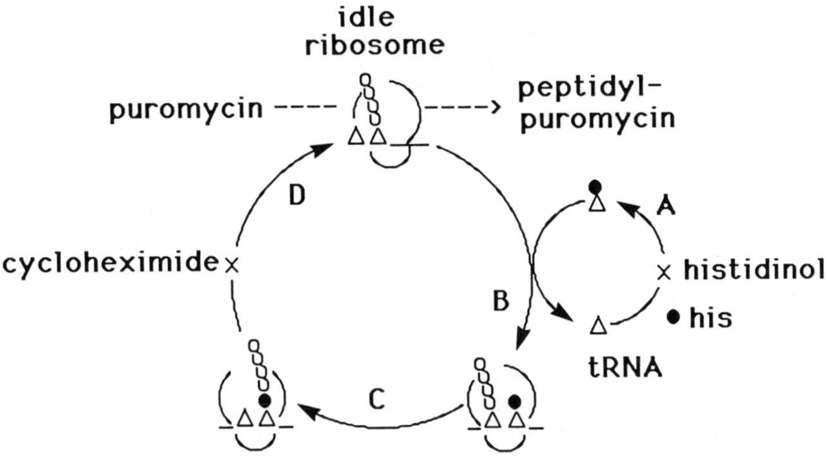

Figure 8: Effects of histidinol and cycloheximide on ribosomal idling. The scheme of the ribosomal elongation cycle includes synthesis of histidyl-tRNA (A), binding of histidyl-tRNA to the ribosomal aminoacyl site (B), synthesis of the peptide bond resulting on the peptidyl-tRNA in the aminoacyl site (C), and translocation of the peptidyl-tRNA to the peptidyl site, which produces a new free aminoacyl site (D). The traditional view that the deacylated tRNA is released during translocation has been changed by recent evidence that it remains associated to a third ribosomal site until binding of the next aminoacyl-tRNA (68). When a ribosome remains in the posttranslocation stage, with an empty aminoacyl site for lack of the corresponding aminoacyl-tRNA, it is defined as "idle ribosome" (33). Puromycin reacts with the peptidyl-tRNA in this configuration. The effects of histidinol and cycloheximide are shown.

sive depletion of his-tRNA is necessary before histidinol affects protein synthesis (Figure 7, top). This is an expected consequence of the slower rate of utilization of his-tRNA for protein synthesis in the presence of cycloheximide. (b) Because we can thus uncouple the effects of histidinol on his-tRNA and ribosomal function, we can now determine whether the regulatory mechanism recognizes directly the levels of his-tRNA or if it is also affected by the functional state of the ribosomes. Ribosomal function does indeed affect the response (Figure 7, bottom). In the presence of 1.5 μg/ml cycloheximide a full response can still be attained, but it requires a more extensive depletion of his-tRNA.

Stimulation of protein degradation in the presence of 1.5 μg/ml cycloheximide is attained only when the depletion of his-tRNA is sufficient to inhibit protein synthesis (Figure 7, top) and to increase the number of ribosomes idling in the his codon with an empty amino acyl site. This we measure by the reactivity of ribosomes *in vivo* to low concentrations of

Table 1: Effects of Cycloheximide and Histidinol on the Synthesis of Peptidyl-^{3}H-puromycin.

	^{3}H-puro-pep %	his idling %
no drugs	100 ± 3.9	3
cycloheximide, 1.5 μg/ml	6.9 ± 0.6	0.2
cycloheximide, 1.5 μg/ml, histidinol, 1 mM.	23.1 ± 0.6	16.2

Cells were incubated for 10 min in the presence of 10^{-6}M ^{3}H-puromycin (1 mCi/mmol). Synthesis of peptidyl-^{3}H-puromycin was determined as trichloroacetic acid insoluble radioactivity in quadruplicate samples and expressed as per cent of the control (*second column*). At this low concentration of puromycin, synthesis of peptidyl-puromycin is linear with incubation time and puromycin concentration; protein synthesis is not affected (at most 1/200 of the ribosomes react with puromycin in the 10 min period). The reactivity to puromycin is taken as a relative measure of idle ribosomes. An estimate of ribosomes idling in the his codon is offered (his-idling, *third column*). In the absence of histidinol, it is assumed that idling in the his codon is proportional to the average his content in cellular protein (3%). In the presence of histidinol, it is assumed that all the additional reactive sites represent ribosomes idling in the his codon. (From Scornik, 1988).

^{3}H-puromycin. Synthesis of peptidyl-puromycin requires a peptidyl tRNA in the ribosomal peptidyl site and an empty amino acyl site, precisely the configuration of idle ribosomes (Figure 8). At the concentration used, 10^{-6}M, puromycin does not significantly affect protein synthesis and can be used as a probe. The reactivity to puromycin is taken as a relative measure of idle ribosomes. A concentration of histidinol sufficient to produce a half-maximal response results in a three-fold increase in the puromycin reactivity and a much larger increase in the estimated number of ribosomes idling in the histidine codon (Table 1).

POSSIBLE NATURE OF THE REGULATORY SIGNAL

The apparent requirement for idle ribosomes brings the mechanism of the mammalian regulation one step closer to that of the stringent response in bacteria. The results seem to rule out mechanisms in which tRNA interacts with other cellular components independently of ribosomes, for instance the production of Ap4A by his-tRNA and its synthetase (66) or the reported activation by tRNA^{His} of the ubiquitin conjugation system (11), a component of the cytosolic pathway of protein degradation (see F). By analogy to the bacterial response we assume that binding of deacylated tRNA to idle ribosomes is a necessary component of the regulation. The recognition of both a deacylated tRNA and ribosomes idling on the corresponding codon has obvious advantages to the cell. In a single step it monitors both the demand for a particular amino acyl-tRNA (idle ribosomes) and its potential supply (the available pool of deacylated tRNA).

The next, unknown, step is probably different from bacteria; ppGpp appears not to occur in mammalian cells (82). This may be due to lack of a stringent factor in mammalian cells because when bacterial stringent factor is added *in vitro* to mammalian ribosomes under appropriate conditions ppGpp is formed (84). It is interesting to speculate why ppGpp is present only in prokaryotes. Bacteria grow so rapidly that instantaneous adaptation to environmental changes is essential. This is most directly accomplished by a one-step regulation in which the molecular signal produced by the stringent factor is the actual effector. At least for the transcriptional component of the stringent response, this is indeed the case; ppGpp inhibits rRNA transcription in purified cell-free systems (see A). *In vivo*, this requires the rapid buildup of fairly high concentrations of ppGpp, 10^{-4} to 10^{-3}M, a process that is facilitated by the enormous concentration of ribosomes in rapidly growing bacteria. In mammalian cells ribosomal concentration is at least two orders of magnitude lower; effective concentrations of ppGpp may be impossible to achieve. Besides, regulation of transcription is more complex in eukaryotes; ppGpp may be useless as an effector. The functional differences between prokaryotes and eukaryotes may be even wider for other components of the response. For instance, the stimulation of protein breakdown in amino acids-starved mammalian cells results from activation of autophagy (see F), a process with no bacterial counterpart.

Because of their slower growth and relatively low concentration of ribosomes it is more likely that in mammalian cells the unknown signal originated in idle ribosomes is the first step of an amplifying sequence of regulatory events. The response does not require synthesis of new proteins because it can be obtained with histidinol in the presence of enough cycloheximide to inhibit protein synthesis by more than 90% (Figure 6). It is also improbable that irreversible inactivation or degradation of a rapidly turned over protein is involved; if it were, replacement of the lost protein would be necessary for restoration of basal rates of protein breakdown; instead, recovery after removal of histidinol can be observed in the presence of cycloheximide (unpublished observations). Finally, it is unlikely that the signal originates in cellular membranes because most polyribosomes (and therefore idle ribosomes) are embedded in the cytoskeleton away from membranes (62,47,24) (with the exception of polyribosomes involved in the synthesis of organelle proteins). The signal originated in idle ribosomes could be a diffusible small molecule (perhaps a regulatory nucleotide different from ppGpp) or a reversibly modified protein. Protein phosphorylation is an obvious candidate, particularly because of the recent report of phosphorylation of translational initiation factor eIF2 after inactivation of leucyl-tRNA synthetase in CHO cells (12). Another possible candidate of protein modification is ADP-ribosylation, because inhibitors of this process prevent the stimulation of protein breakdown by histidinol (73).

TWO MECHANISMS FOR CELLULAR PROTEIN DEGRADATION

Our research is currently focused on the nature of the regulatory signals initiated by deacylation of tRNA and ribosome idling. A full elucidation of the process should also provide an insight into how RNA transcription and translational initiation are inhibited, or protein breakdown stimulated. Of these, we may be closest to understanding the regulation of polypeptide chain initiation, where inhibition is possibly mediated by phosphorylation of one of the initiation factors, eIF2 (13). Attempts to understand the molecular basis of the stimulated protein degradation may be at present premature because our knowledge of the mechanisms by which cells degrade their proteins is still very limited. Two pathways are known. One is the sequestration of cytoplasmic components in autophagic vacuoles, where they are degraded by lysosomal enzymes (16,54,49). This mechanism can be directly visualized by electron microscopy. Autophagy may occur preferentially in elements of the vacuolar apparatus adjacent to Golgi cisternae (55). The abundance of autophagic vacuoles in rat hepatocytes parallels the protein degradative activity of these cells, under a variety of experimental conditions (52). Other evidence for the occurrence of autophagy are the accumulation of glycogen (a typical cytosolic component) within lysosomes in the liver of children with a genetic deficiency of lysosomal alpha-glucosidase (glycogenosis type II) (34), the finding of partially degraded protein in lysosomes isolated from rat liver (52,45,29) and muscle (28), the sequestration in vacuoles of fluorescent or radioactive proteins and non-permeable sugars introduced in cells by microinjection (83,17,80), and the inhibition of the degradation of cellular proteins by inhibitors of lysosomal cathepsins, such as leupeptin (45) and pepstatin (14), or by permeable weak bases that interfere with acidification of the intralysosomal space, such as chloroquine (97,15).

The second pathway for the degradation of cellular proteins is the direct attack by proteolytic enzymes present in the cytosol. Reticulocytes, in which lysosomes are rare, degrade newly synthesized hemoglobin very rapidly when the globin is abnormal, either because of a genetic defect or because of the introduction of amino acid analogs. Cytosolic reticulocyte extracts contain an ATP-dependent proteolytic system which can selectively and rapidly degrade abnormal and denatured globin as well as other proteins (18,36,35,67). The first step in this degradation is the formation of covalent conjugates of the substrate protein with one or more molecules of ubiquitin, a small protein. A mutant cell line with a temperature-sensitive defect in the ubiquitin-conjugation system does not degrade short-lived proteins (10). It seems clear that this cytosolic pathway plays a major role in the selective degradation of abnormal and short-lived proteins. The extent to which cytosolic enzymes (ubiquitin-dependent or not) are responsible for the degradation of more stable cellular proteins remains at this time obscure.

Stimulation of protein degradation by amino acid deprivation seems to be caused by increased autophagy. Quantitative correlation between autophagy and stimulated protein breakdown has been well established in liver (52). In other mammalian cells, the evidence rests largely on the sensitivity of stimulated protein degradation to inhibitors of lysosome function (44,1,81). In CHO cells, stimulated (but not basal) degradation is sensitive to leupeptin and chloroquine (72). It has been reported that ribonuclease treatment inactivates the ubiquitin-conjugation system in reticulocytes; the activity is restored by addition of $tRNA^{His}$, an effect superficially reminiscent of the regulatory phenomenon discussed here (11). It is, however, improbable that this requirement explains the stimulation of protein breakdown in amino acid-deprived cells: stimulated degradation is in all likelihood autophagic, the report did not indicate whether deacylation of tRNA is essential for the effect and ribosomes were not present in the purified cell-free system used. More recently, the requirement for tRNA was shown for some protein substrates but not others (20). It was suggested that susceptibility to degradation by the cytosolic system depends on the nature of the N-terminal amino acid and that for some proteins a tRNA-dependent addition of an amino acid to the amino terminus of the substrate may trigger its degradation (3).

Understanding the molecular basis of stimulated autophagy may have to wait until this process can be studied in cell-free systems. Autophagic vacuoles containing sequestered proteins and degradative enzymes can be isolated and degradation of the contained substrate proteins carried to completion *in vitro* (29), but autophagic sequestration (let alone its stimulation) has yet to be reproduced in the test tube.

SIGNIFICANCE OF THE REGULATORY RESPONSE TO AMINO ACID DEPRIVATION IN MAMMALIAN CELLS

The adaptive advantages of some of the regulatory changes caused by amino acid deprivation are direct and immediate. When amino acid pools are depleted slower protein synthesis and faster protein breakdown tends to replenish them. Slower synthesis of rRNA reflects the fact that the existing protein synthetic machinery exceeds the availability of amino acid precursors. The overall physiological role of the response, however, may depend on the functional context in which it occurs.

Growth Control

Growing cells deprived of one or more essential amino acids often become quiescent. They cease to accumulate protein and RNA; rates of protein synthesis decrease and those of protein degradation increase until both rates become equal. Total protein content tends to remain constant for prolonged periods and the cells remain viable. Growth is arrested in the same stage of the cell cycle as after deprivation of serum or growth factors (60). Histidinol has the same effect (94). It has been proposed that the effects of serum factors and amino acids on growth is mediated by a common pleiotypic mechanism (37). It may even be that the effects of some growth factors are mediated by changes in the intracellular concentration of amino acids and other nutrients (39). The regulatory mechanism discussed in this article may be related to the pleiotypic control of growth, but it is not necessarily coupled to growth control. In CHO cells, serum or amino acid deprivation stimulate protein breakdown but their effects are independent and additive (76). At concentrations sufficient to build up a fully charged tRNA, cycloheximide cancels the effect of histidine starvation or histidinol but not those of serum deprivation (unpublished results). Furthermore, amino acid deprivation can induce stimulation of protein degradation and polyribosome disaggregation in non-growing cells, such as hepatocytes (50,79,21).

Protection from Translational Errors

In the absence of an amino acyl tRNA, idle ribosomes tend to accept a tRNA carrying a different amino acid. High concentrations of histidinol, for instance, result in abnormal proteins in which glutamine replaces histidine (59). An increased frequency of translational errors could have catastrophic consequences for the cell (57). Replenishment of intracellular amino acids by increased protein degradation and inhibition of translational initiation can prevent the synthesis of abnormal proteins. This role is well established in bacteria by the large increase in mistranslation in relaxed mutants deprived of a needed amino acid (26).

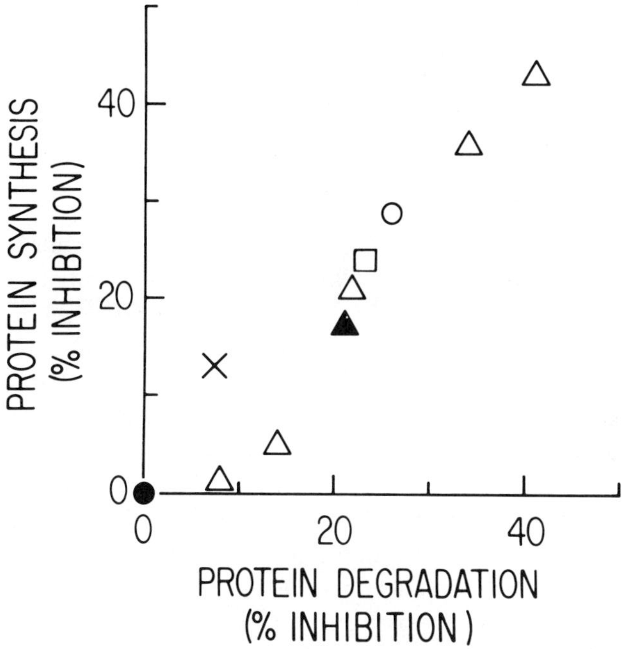

Figure 9: Inhibition by chloroquine and leupeptin of protein degradation (*abscissa*) and synthesis (*ordinate*) as % of control values under the same conditions without drugs. *Open triangles*: Chloroquine 0.4, 1.2, 4, 12 and 40 μM (no amino acids, no serum). *Open circle*: 40 μM (no amino acids). *Open square*: 40 μM (no serum). Cross: 40 μM (complete). *Filled triangle*: Leupeptin, 0.45 mg/ml (no serum, no amino acids). (From 72).

Maintenance of Intracellular Amino Acid Pools

The most obvious role of stimulated protein degradation in amino acid-deprived cells is the replenishment of intracellular pools. In the absence of extracellular amino acids protein degradation is the only source of essential amino acids, as well as nitrogen available for the synthesis of the non-essential ones. Faster degradation of existing proteins permits the continuing synthesis of new protein molecules. This function permits the replacement of short-lived or secreted proteins, as well as induced enzymes required for the adaptation to the new metabolic condition at the expense of relatively abundant long-lived constitutive cellular components. Because of this protein turnover, CHO cells can maintain about half their normal rate of protein synthesis in the absence of amino acids in the medium (72). In this situation, inhibition of stimulated protein breakdown by chloroquine or leupeptin produces a proportional decrease in the rate of protein synthesis (Figure 9).

Lower protein synthesis is largely due in these cases to reduced translational initiation (see B). As we have seen above, this control may avoid mistranslation. It is also possible that a general inhibition of initiation favors the preferential translation of some mRNAs; to our knowledge, this possible role of the regulatory response has not yet been explored.

Regulation of Extracellular Amino Acid Pools

The functions of the regulatory response discussed so far are common to unicellular organisms and cells of higher organisms. In multicellular organisms, changes in the relative rates of protein synthesis and degradation serve also to regulate extracellular pools. Bacteria deprived of required amino acids or a nitrogen source reutilize the amino acids derived from protein breakdown very efficiently, with little loss to the environment (48). In mammalian cells, when breakdown exceeds synthesis most of the unutilized amino acids are exported. In amino acid-deprived CHO cells, two-thirds of the amino acids produced by protein breakdown are released to the medium; if the stimulated protein degradation is inhibited by chloroquine, both protein synthesis and amino acid export are affected (72). In monolayer cultures, where the volume of medium is at least three orders of magnitude larger than the volume of cells, the released amino acids are diluted away and wasted. In intact animals, however, the phenomenon plays a major physiological role. The amino acids accumulate in the extracellular fluid where they remain available to the cells or are transported by the circulation to other tissues (96,71).

Feeding in most adult mammals is discontinuous, and dietary amino acids are stored as excess body proteins. Changes in opposite directions of protein synthesis and breakdown permit net storage of amino acids after a protein-rich meal and net release between meals (27,42,96,6,75). Tissues with fast turnover and ample responses tend to store more than their share, and export amino acids readily for use by more conservative tissues. These changes in protein metabolism in non-growing differentiated tissues can be induced by hormones, notably insulin and (in liver) glucagon and by the amino acids themselves (50), but we do not fully know at this stage the relative importance of these regulators in the intact animal or which part deacylation of tRNA plays in these effects.

Some effects of amino acids on protein metabolism in isolated cells and tissues can be observed in the presence of cycloheximide and must be therefore independent of the utilization of the amino acids for protein synthesis (see C). We do not know whether these cycloheximide-insensitive effects are physiologically important or not. Different regulatory mechanisms may respond to different needs. The net effect of the tRNA-dependent regulation described in this article is to minimize the effects of starvation and adjust the rates of protein synthesis (and hence growth) to the supply of amino acids. This mechanism operates when the intracellular amino acid concentrations are at or below the K_m of the corresponding tRNA synthetases (in the order of 10^{-6} to 10^{-5}M). On the other hand, some of the effects of amino acids in liver or muscle may respond to the need of the whole animal to store (rather than catabolize) amino acids when supplied in excess by the diet, under conditions when the concentration of amino acids may be much higher.

CONCLUSIONS

Mammalian cells respond to amino acid deprivation with polyribosome disaggregation, inhibition of rRNA synthesis and faster protein degradation. These physiological changes adjust the rate of growth to the availability of amino acids, prevent translational errors and restore intra- and extracellular amino acid pools. At least in one case, stimulation of protein breakdown by histidine deprivation of CHO cells, the regulatory mechanism is linked to the utilization of the amino acid for protein synthesis. It recognizes the steady-state levels of histidyl-tRNA and the appearance of idle ribosomes. The mechanism is similar to that of the bacterial stringent response, except for the absence of ppGpp in mammalian cells. The increased rate of protein degradation is probably due to increased autophagy. This study is still in its infancy; its major challenges are the identification of the regulatory signal originating in idle ribosomes, the molecular mechanism of stimulated autophagy, and the events linking the signal to this and the other components of the regulatory response. Our major limitation is the lack of a cell-free system in which the regulation can be reproduced. The field has attracted so far only a small group of persistent researchers. We hope that the elucidation of the initial events in the regulation, described here, increased awareness of the central physiological role of this phenomenon and advances in related areas of cell biology will encourage others to help meet those challenges.

ACKNOWLEDGEMENT

This work is supported by NSF grant DMB88-03517.

REFERENCES

1. Amenta, J. S., Sargus, M. J., and Baccino, F. M. *Biochem. J.*, **168**:223-227, (1977).
2. Austin, S. A., Pain, V. M., Lewis, J. A., and Clemens, M. J. *Eur. J. Biochem.*, **122**:519-526, (1982).
3. Bachmair, A., Finley, D., and Varshavsky, A. *Science*, **234**:179-186, (1986).
4. Baxter G. C., Stanners, C. P. J. *Cell Physiol.*, **96**:139-146, (1978).

5. Borek, E., Ryan, A, Rockenbach, J. *J. Bacteriol.*, **69**:460-467, (1955).
6. Botbol, V., Scornik, O. A. **In**: *Intracellular Protein Catabolism*, E. A. Khairallah, J. S. Bond, J. W. C. Bird, (Eds.) New York:Alan Liss, pp. 573-583, (1985).
7. Buse, M. G., Biggers, J. F., Friderici, K. H., Buse, J. F. *J. Biol. Chem.*, **247**:8085-8096, (1972).
8. Cashel, M. *Annu. Rev. Microbiol.*, **29**:301-318, (1975).
9. Cashel, M., Rudd, K. E. **In**: Escherichia coli *and* Salmonella typhimurium. *Cellular and Molecular Biology*. Neidhardt, F. C. (ed.). Washington: American Society for Microbiology, pp. 1410-1438 (1987).
10. Ciechanover, A., Finley, D., Varshavsky, A. *Cell*, **37**:57-66, (1984).
11. Ciechanover, A., Wolin, S. L., Steitz, J. A., Lodish, H. F. *Proc. Natl. Acad. Sci. USA*, **82**:1341-1345, (1985).
12. Clark, J. L., Rabe, J., Arfin, S. M. *J. Cell Physiol.*, **98**:237-240, (1979).
13. Clemens, M. J., Galpine, A., Austin, S. A., Panniers, R., Henshaw, E. C., Duncan R., Hershey, J. W. B., Pollard J. W. *J. Biol. Chem.*, **262**:767-771, (1987).
14. Dean, R. T. (1975) *Nature*, **257**:414-416, (1975).
15. Dean, R. T., Jessup, W., Roberts, C. R. *Biochem. J.*, **217**:27-40, (1984).
16. de Duve, C., Wattiaux, R. *Annu. Rev. Physiol.*, **28**:435-492, (1966).
17. Dice, J. F., Backer, J. M., Miao, P., Bourret, L., McElligott, M. A. **In**: *Intracellular Protein Catabolism*. E. A. Khairallah, J. S. Bond, J. W. C. Bird, (Eds.) New York:Alan Liss, pp. 385-394, (1985).
18. Etlinger, J. D., Goldberg, A. L. *Proc. Natl. Acad. Sci. USA*, **74**:54-58, (1977).
19. Fan, K., Fisher, K. M., Edlin, G. *Exp. Cell Res.*, **82**:111-118, (1973).
20. Ferber S., Ciechanover, A *J. Biol. Chem.*, **261**:3128-3134, (1986).
21. Flaim, K. E., Peavy, D. E., Everson, W. V., Jefferson. L. S. *J. Biol. Chem.*, **257**:2932-2938, (1982).
22. Franze-Fernández, M. T., Pogo, A. O. *Proc. Natl. Acad. Sci. USA*, **68**:3040-3044.
23. Franze-Fernández, M. T., Fontanive-Sangüesa A. V. *Biochem. Biophys. Acta.*, **331**:71-80, (1973).
24. Fulton, A. B., Wan, K. M., Penman, S, *Cell*, **20**:849-857, (1980).
25. Gallant, J. A. Annu. Rev. Genet., **13**:393-415, (1979).
26. Gallant, J. A., Foley, D. **In**: *Regulation of Macromolecular Synthesis by Low Molecular Weight Mediators*. G. Koch, D. Richter, (Eds.) New York:Academic, pp. 5-23, (1979).
27. Garlick, P. J., Millward, D. J., James, W. P. T. *Biochem. J.*, **136**:935-945, (1973).
28. Gerard, K. W., Schneider, D. L. *J. Biol. Chem.*, **254**:11798-11805, (1979).
29. Glaumann, H., Ahlberg, J., Berkenstam, A., Henell, F. **In**: *Intracellular Protein Catabolism*. E. A. Khairallah, J. S. Bond, J. W. C. Bird, (Eds.) New York:Alan Liss, pp. 417-426, (1985).
30. Goldberg, A. L., St. John, A. C. *Annu. Rev. Biochem.*, **45**:747-803, (1976).
31. Grummt, F., Grummt I., *Eur. J. Biochem.*, **64**:307-312, (1976).
32. Gunn, J. M., *Exp. Cell. Res.*, **117**:448-451, (1978).
33. Haseltine, W. A., Block, R. *Proc. Natl. Acad. Sci. USA*, **70**:1564-1568, (1973).
34. Hers, H. G., deBarsy, T. In: *Lysosomes and Storage Diseases* H. G. Hers, van Hoff (Eds.) New York:Academic, pp. 197-216, (1973).
35. Hershko, A. *J. Biol. Chem.* **263**: 15237-15240 (1988)
36. Hershko, A., Ciechanover, A. *Annu. Rev. Biochem.*, **51**:335-364, (1982).
37. Hershko, A., Mamont, P., Shields, R., Tomkins, G. M. *Nature* New Biol., **232**:206-211, (1971).
38. Hogan, B. L. M., Korner, A. *Biochem. Biophys. Acta.*, **169**:129-138, (1968).
39. Holley, R. W., Kiernan, J. A. *Proc. Natl. Acad. Sci. USA*, **71**:2942-2945, (1974).
40. Hopgood, M. F., Clark, M. G., Ballard, F. J. (1977) *Biochem. J.*, **164**:399-407, (1977).
41. Kajitani, M., Ishihama, A. *J. Biol. Chem.*, **259**:1951-1957, (1984).
42. Khairallah, E. A. **In**: *Protein Turnover and Lysosomal Function*. H. L. Segal, D. J. Doyle, (Eds.) New York:Academic, pp. 89-104, (1978).

43. Khairallah, E. A., Mortimore, G. E. *J. Biol. Chem.*, **251**:1375-1384, (1976).
44. Knowles S. E., Ballard F. J. *Biochem. J.*, **156**:609-617, (1976).
45. Kominami E., Hashida, S., Khairallah, E. A., Katunuma, N. *J. Biol. Chem.*, **258**:6093-6100, (1983).
46. Laffler T., Gallant J. A. *Cell*, **3**:47-49, (1974).
47. Lenk, R., Ransom, L., Kaufmann, Y., Penman, S. *Cell*, **10**:67-78, (1977).
48. Mandelstam, J. *Biochem. J.*, **69**:103-110, (1958).
49. Marzella L., Glaumann, H. **In**: *Lysosomes: Their Role in Protein Breakdown*. H. Glaumann, F. J. Ballard (Eds.) New York: Academic, pp.319-367, (1986).
50. Mortimore, G. E. **In**: *Lysosomes: Their Role in Protein Breakdown* H. Glaumann, F. J. Ballard (Eds.) New York: Academic, pp.415-443, (1986).
51. Mortimore, G. E., Ward, W. F. **In**: *Lysosomes in Biology and Pathology*. J. T. Dingle, R. T. Dean (Eds.) Amsterdam:North Holland, vol.5, pp. 157-184, (1976).
52. Mortimore, G. E. Ward, W. F. Schworer, C. M. **In**: *Protein Turnover and Lysosomal Function*. H. L. Segal, D. J. Doyle, (Eds.) New York:Academic, pp. 67-87, (1978).
53. Neidhardt, F. C. *Bact. Rev.*, **30**:701-719, (1966).
54. Novikoff, A. B. **In**: *Lysosomes and Storage Diseases* H. G. Hers, van Hoff (Eds.) New York:Academic, pp. 1-41, (1973).
55. Novikoff, A. B. *Proc. Natl. Acad. Sci. USA*, **73**:2781-2787, (1976).
56. Ogilvie, A. Huschka, V. Kersten, W. *Biochem. Biophys. Acta.*, **565**:293-304, (1979).
57. Orgel, L. E. *Proc. Natl. Acad. Sci. USA*, **49**:517-521, (1963).
58. Pao, C. C. Dennis, P. P. Gallant, J. A. *J. Biol. Chem.*, **255**:1830-1833, (1980).
59. Parker, J. Pollard, J. W. Friesen, J. D. Stanners, C. P. *Proc. Natl. Acad. Sci. USA*, **75**:1091-1095, (1978).
60. Pardee, A. B. Dubrow, R. Hamlin, J. L. Kletzien, R. F. *Annu. Rev. Biochem.*, **47**:715-750, (1978).
61. Pollard, J. W., Galpine, A. R., Clemens, M. J. *Eur. J. Biochem.* **182**: 1-9 (1989).
62. Porter, K. R. **In**: *Cell Motility* R. Goldman, T. Pollard, J. Rosenbaum (Eds.), New York:Cold Spring Harbor, pp.1-28, (1976).
63. Pösö, A. R., Mortimore, G.E. *Proc. Natl. Acad. Sci.*, **81**:4270-4274, (1984).
64. Pösö, A. R. Wert, J. J. Mortimore, G. E. *J. Biol. Chem.*, **257**:12114-12120, (1982).
65. Pronczuk, A. W. Rogers, Q. R. Munro, H. N. *J. Nutr.*, **100**:1249-1258, (1970).
66. Rapaport, E., Zamecnick, P. C. *Proc. Natl. Acad. Sci. USA*, **73**:3984-3988, (1976).
67. Rechsteiner, M. (ed.) *Ubiquitin*. New York: Plenum, (1988).
68. Rheinberger, H. J., Sternbach, H., Nierhaus, K. H. *J. Biol. Chem.* **261**: 9140-9143 (1986).
69. Sarmientos, P., Sylvester, J. E., Costente, S. Cashel, M. *Cell*, **32**:1337-1346, (1983).
70. Scornik, O. A. *J. Biol. Chem.*, **258**:882-886, (1983).
71. Scornik, O. A. *Federation Proc.*, **43**:1283-1288, (1984a).
72. Scornik, O. A. *J. Cell. Physiol.*, **121**:257-262, (1984b).
73. Scornik, O. A. *Fed. Proc.* **46**: 2086 (abs.) (1987).
74. Scornik, O. A. *J. Cell Biol.* **103:187a. (and** *ms. in preparation*).
75. Scornik, O. A., Botbol, V. **In**: *Lysosomes: Their Role in Protein Breakdown*. H. Glaumann, F. J. Ballard (Eds.) New York: Academic, pp.445-484, (1986).
76. Scornik, O. A., Ledbetter, M. L. S., Malter, J. S. *J. Biol. Chem.*, **255**:6322-6329, (1980).
77. Scornik, O. A., Schadel, B. F., Franze-Fernndez, M. T., Botbol, V. *Federation Proc.*, **37**:1332, (1978).
78. Seglen, P. O. **In**: *Protein Turnover and Lysosomal Function* H. L. Segal, D. J. Doyle, (Eds.) New York:Academic, pp.431-453, (1978).
79. Seglen, P. O. **In**: *Lysosomes: Their Role in Protein Breakdown*. H. Glaumann, F. J. Ballard (Eds.) New York: Academic, pp.371-414, (1986).
80. Seglen, P. O., Gordon, P. G., Tolleshaug, H., Hoyvik, H. **In**: *Intracellular Protein Catabolism*. E. A. Khairallah, J. S. Bond, J. W. C. Bird, (Eds.) New York:Alan Liss, pp. 437-446, (1985).

81. Seglen, P. O., Solheim, A. E., Grinde, B., Gordon, P. B., Schwarze, P. E., Gjessing, R., Poli, A. *Ann. N.Y. Acad. Sci.*, **349**:1-17, (1980).
82. Silverman, R., Atherly, A.; Richter, D. **In**: *Regulation of Macromolecular Synthesis by Low Molecular Weight Mediators*. G. Koch, D. Richter, (Eds.) New York:Academic, pp.115-125, (1979).
83. Stacey, D. W., Allfrey, V. G. *J. Cell. Biol.*, **75**:807-817, (1977).
84. Stanners, C. P. **In**: *Growth Kinetic and Biochemical Regulation of Normal and Malignant Cells*. B. Drewinko, R. M. Humphrey, (Eds.) Baltimore:Williams & Wilkins, pp. 35-47, (1977).
85. Stanners, C. P., Thompson, L. H. **In**: *Control of Mammalian Cell Proliferation* B. Clarkson, R. Baserga (Eds.) New York:Cold Spring Harbor, pp.191-203, (1974).
86. Stamminger, G., Lazzarini, R. *Cell*, **1**:85-90, (1974).
87. Stent, G. S., Brenner, S. *Proc. Natl. Acad. Sci. USA*, **47**:2005-2014, (1961).
88. St. John, A. C., Conklin, K., Rosenthal, E., Goldberg, A. L. *J. Biol. Chem.*, **253**:3945-3951, (1978).
89. Thompson, L. H., Lofgren, D. J., Adair, G. M. *Somatic Cell Genet.*, **4**:423-435, (1978).
90. Tischler, M. E., Desautels, M., Goldberg, A. L. *J. Biol. Chem.*, **257**:1613-1621, (1982).
91. Van Venrooij, W. J. W, Henshaw, E. C., Hirsch, C. A. *Biochem. Biophys. Acta.*, **259**:127-137, (1972).
92. Vaughan, M. H., Hansen, B. S. *J. Biol. Chem.*, **248**:7087-7096, (1973).
93. Warburton, M. J., Poole, B. *Proc. Natl. Acad. Sci. USA*, **74**:2427-2431, (1977).
94. Warrington, R. C., Hechtman, R. (1977) *Cell Biol. Int. Reports*, **1**:571-579, (1977).
95. Warrington, R. C., Wratten, N., Hechtman, R. *J. Biol. Chem.* **252**:5251-5257, (1977).
96. Waterlow, J. C., Garlick, P. J., Millward, D. J. *Protein Turnover in Mammalian Tissues and in the Whole Body*. Amsterdam:North Holland, (1978).
97. Wibo, M., Poole, B. *J. Cell Biol.*, **63**:430-440, (1974).
98. Woodside, K. H., Mortimore, G. E. *J. Biol. Chem.*, **247**:6474-6481, (1972).

7

BIOCHEMICAL AND PHYSIOLOGICAL FUNCTIONS OF PARASITE PROTEINASES

Marc H. Dresden[1]

Verna and Marrs McLean Department of Biochemistry
Baylor College of Medicine
Houston, TX 77030

1 Deceased February 17, 1990

INTRODUCTION

The functions of proteolytic enzymes in biological systems are of increasing interest in modern biology and biochemistry. Such well-known enzyme systems as the pancreatic proteinases involved in the digestion of protein foods, the cascade systems of blood coagulation and the complement system have demonstrated specific roles for generally well-characterized proteinases. Interesting but less well-documented roles for proteolytic enzymes have been suggested for collagenases in tumor invasion, wound healing, and uterine involution, for acrosomal proteinases in sperm penetration of the mammalian egg, and for leucocyte proteinases in inflammatory reactions, *etc.*

In the case of parasitic organisms, relatively little information exists both for the types and variety of proteolytic enzymes and their functions. Proteinases undoubtedly serve general functions such as cellular turnover of proteins, *etc.*, in parasites. In addition, one might surmise that proteolytic enzymes would serve three specific functions unique to parasites. The first reflects the invasion of host tissues and/or cells as part of the parasite life cycle. For example, trematodes such as *Schistosoma mansoni* or *Fasciola hepatica* infect their vertebrate hosts by penetration through host skin and/or intestinal tissues. Other like stages, for example, the miracidia of *S. mansoni* invade their intermediate, invertebrate hosts by penetration through connective tissues. *Taenia*, a cestode, invades muscle. Some nematodes, such as *Strongyloides* and *Ascaris*, also penetrate host tissues. In addition, intracellular parasites, including parasites causing malaria, babesiosis, leishmaniasis, sleeping sickness, *etc.*, must invade host circulatory cells including red blood cells and macrophages.

A second role of proteolytic enzymes in parasites revolves around the nutritional needs of the parasite. While some parasites are able to transport nutrients efficiently through their outer membranes or surfaces, many extracellular and intracellular parasites utilize host proteins and their constituent amino acids for nutrition. Proteolytic enzymes capable of hydrolyzing host proteins including blood proteins such as hemoglobin and albumin would clearly be useful in this regard.

A further role for proteolytic enzymes might occur in the transformation of one stage of a parasite's life cycle to another stage. This kind of role for proteinases has been elegantly demonstrated by Kafatos in a series of studies of the role of the serine, trypsin-like proteinase, cocoonase, in the escape of the silkworm from its cocoon. Analogous situations may occur in the hatching of trematode miracidia from its egg shell, the loss of cuticle from infective nematode larvae, *etc.*

While this chapter is not meant to be an exhaustive review of the literature, I will address examples of the first two of these postulated roles for parasite proteinases.

NUTRITIONAL FUNCTION OF PROTEINASES

Mammals, which contain extensive open-ended digestive tracts, have been shown to contain a wide variety of proteolytic enzymes involved in digestion of protein foodstuffs. These include acid pH-optima enzymes, such as pepsin and renin of the primary stomach, and alkaline proteinases such as elastase, trypsin, chymotrypsin and a variety of exopeptidases produced by the pancreas and intestinal mucosa.

Many extracellular parasites, such as the trematodes, contain blind intestinal tracts so that undigested foodstuffs are regurgitated *via* the same channels used for intake. A number of these parasites are known to take in host blood and tissue proteins, which are utilized in nutrition of the parasite. Enzymes able to digest hemoglobin have been described for the trematode parasites, *Schistosoma mansoni* (1), as well as other schistosome species (2-7), *Fasciola hepatica* (8-11), and *Clonorchis sinensis* (12). Maki *et al.* (13) have described hemoglobinase activity in cestodes, including *Diphyllobothrium erinacei* and *Hymenolepis nana* as well as nematodes, including *Angiostrongylus cantonensis, Dirofilaria immitis, Trichuris muris* and *Ascaris suum.* Further, Lumley and Lee (14) and Oya and Noguchi (15) showed acidic proteinases active against hemoglobin in the parasitic nematodes, *Nippostrongylus braziliensis, Nematodirus battus*, and *Ancylostoma caninum.*

While proteinases able to digest hemoglobin appear to be widely distributed among the multicellular parasites, similar enzymes have also been described in unicellular protozoan parasites. For example, thiol-dependent proteinases have been demonstrated in *Entamoeba histolytica* (17-20), in the flagellate protozoan *Tritichomonas foetus* (21), and in African trypanosomes (22).

The nature of the endopeptidases involved in hemoglobin digestion is not entirely clear, although those described have an acidic pH optimum. The endopeptidase classes which exhibit acid pH optima are the cysteine and aspartate proteinases. Bolla and Weinstein (23) suggested that the hemoglobinolytic activity of *Nippostrongylus braziliensis* was due to a cathepsin D-like aspartate proteinase. A similar suggestion has been made by Bogitsh (24) for *Schistosoma mansoni*. Furthermore, inhibitor studies (10) have revealed that *Fasciola hepatica* uses a similar proteinase in hemoglobin digestion. This activity may be similar to the acidic pH-optimum enzymes reported by Simpkin *et al.* (12) and Locatelli and Paoletti (9). Maki *et al.* (15) have suggested that the acidic, hemoglobin-digesting enzymes found in seven species of parasites (including four nematodes, one trematode, and two cestodes) are due to aspartyl proteinases related to cathepsin D.

It should be pointed out that the term "hemoglobinase" used by many researchers may be a misnomer. Hemoglobin, as a proteinase substrate, can be hydrolyzed by a wide variety of proteinases, including representatives of all four classes of proteolytic enzymes. It is thus unlikely (although not proven) that all such "hemoglobinases" are specific for hemoglobin. On the other hand, it has been clearly shown that a number of these proteinases are very inactive against other blood proteins, such as serum albumin (7). Below we describe the properties and function of cysteine proteinases involved in hemoglobin digestion in *Schistosoma mansoni*.

HEMOGLOBINASE ACTIVITY IN *SCHISTOSOMA*

The human trematode parasite *Schistosoma mansoni* resides in the mesenteric veins of its vertebrate host. It is exposed to the rich nutritional content of the portal circulation and has been shown to contain a number of active transport systems for amino acids and sugars. In addition, the adult worm ingests whole blood from the portal circulation (25). In the worm digestive tract, these red blood cells are lysed and hemoglobin is released. Proteolytic degradation of hemoglobin was first reported in *S. mansoni* worm homogenates by Timms and Bueding (1) and later by Senft and his collaborators (2,3). The contents of the intestinal tract are exchanged every 3 to

4 hours, thus assuring that both undigested host blood components and parasite digestive enzymes are regurgitated into the host circulation.

In previous studies we have shown that *S. mansoni* hemoglobinase activity is stimulated by thiols *in vitro* and belongs to the class of cysteine proteinases (5). Proteinase activity is associated with proteins in a molecular weight range of 28,000-32,000 and has been detected directly in material collected from the worm digestive tract (6). Most recently, two cysteine proteinases have been purified to homogeneity from *S. mansoni* worm extracts (26). One of these proteinases has been designated SMw32-P.

The role of this enzyme in parasite nutrition was tested by examining its action *in vitro* on both hemoglobin and albumin (representing the two major proteins present in blood). As can be seen in Figure 1, the enzyme actively degrades hemoglobin as detected by acrylamide gel electrophoresis. A 21,000-dalton degradation fragment appears during the degradation process. Further experiments using hemoglobin from permissive (human) or non-permissive (bovine, rat) hosts for *S. mansoni* also resulted in rapid degradation of substrate into small peptide products (data not shown). We have now begun to analyze the degradation products with high pressure liquid chromatography and find a definite series of peptides produced by the action of the schistosome enzyme on hemoglobin. Moreover, similar cysteine proteinases from other trematode parasites, *Clonorchis sinensis* and *Fasciola hepatica*, appear to give a different series of peptides. This area is of considerable interest since the cleavage specificities of the cysteine proteinases on protein substrates has not been extensively investigated.

In contrast to hemoglobin, bovine serum albumin, in its undenaturated form, is almost totally resistant to the schistosome proteinase (Figure 2). However, when albumin was denatured prior to assay, the *S. mansoni* proteinase was able to effect a limited hydrolysis that resulted in the accumulation of two polypeptide products with apparent molecular weights of 54,000 and 18,000-20,000. Human serum albumin as substrate yielded the same results. Both hemoglobin and denatured albumin digestion are completely blocked by a specific inhibitor of the cysteine proteinases, E-64 (27,28).

A further point of interest about these proteinases involves their *in vivo* function and regulation (29). The 32,000-dalton *S. mansoni* proteinase is isolated from the parasite in an inactive form and requires thiol activation for *in vitro* assays. We have surmised that the proteinase secreted into the extracellular compartment of the worm gut must also require activation before it can degrade the *in vivo* substrate, hemoglobin. This scenario requires a source of thiols at a specific time and place, *i.e.*, when hemoglobin is present

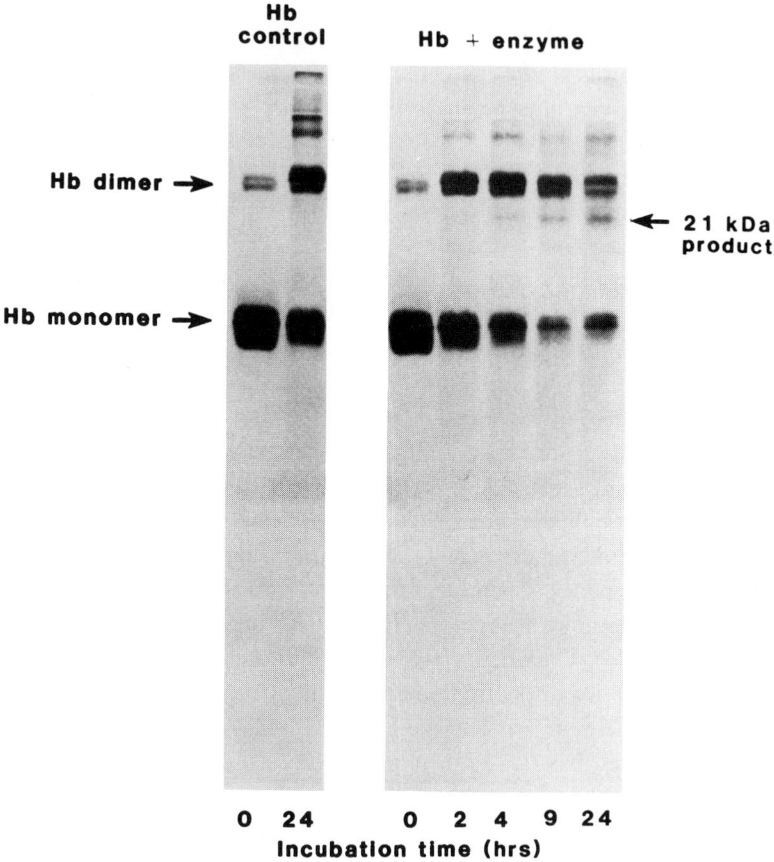

Figure 1: Degradation of hemoglobin (Hb) by the schistosome proteinase, SMw32-P. Hemoglobin, incubated alone (left) or with SMw32-P (right), was visualized by silver stain of a 20% SDS-PAGE. Hb dimer and monomer bands, as well as a Hb degradation product, are indicated by arrows. (reprinted with permission)

within the blind gut of the parasite. The most logical candidate is glutathione, present in abundance in host red blood cells.

Regulation of proteinase activity by thiols was investigated by measuring activation of proteolytic activity over time with different glutathione (GSH) concentration (Figure 3). GSH levels chosen for these experiments included those found in serum (0.02-0.1 mM) and in whole blood (3-3.2 mM) (29).

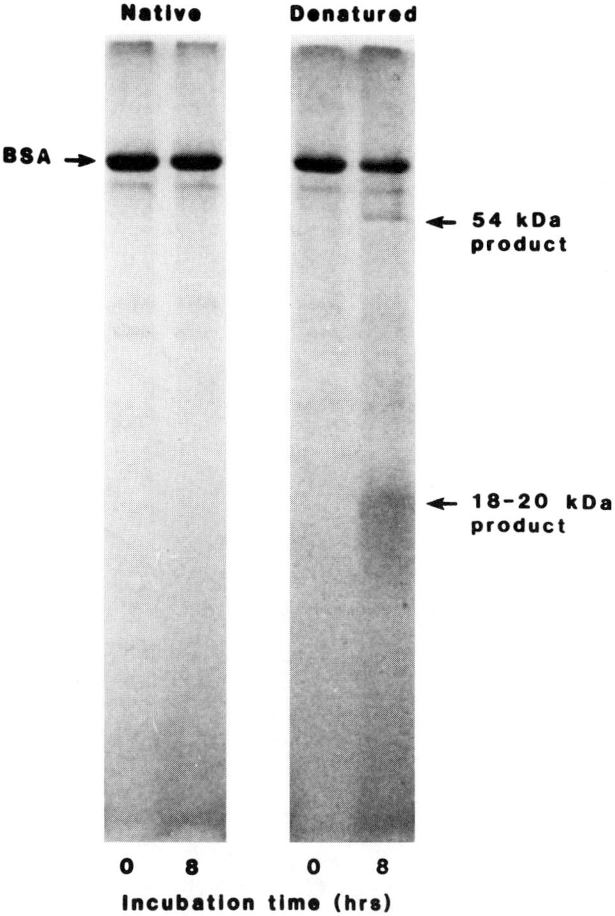

Figure 2: Incubation of serum albumin (bovine) with the SMw32-P proteinase. Native or heat-denatured albumin was assayed in 0.1 M citrate buffer containing 5 mM cysteine. Aliquots of each assay were visualized by silver stain of a 15% SDS-PAGE. Degradation products are indicated by arrows. (reprinted with permission)

Maximal activity was at or above 2 mM glutathione. However, maximal activation could be obtained at lower concentrations when the time of activation was prolonged. This relationship is described further (Figure 4) where proteinase activation was determined as a function of GSH concentration and time. The data are presented as $100(A - A_{Ct})/A$, where A_{Ct} represents the proteinase activity at a particular glutathione concentration and time. When proteinase activity is expressed in these terms, the curve decreases as maximal

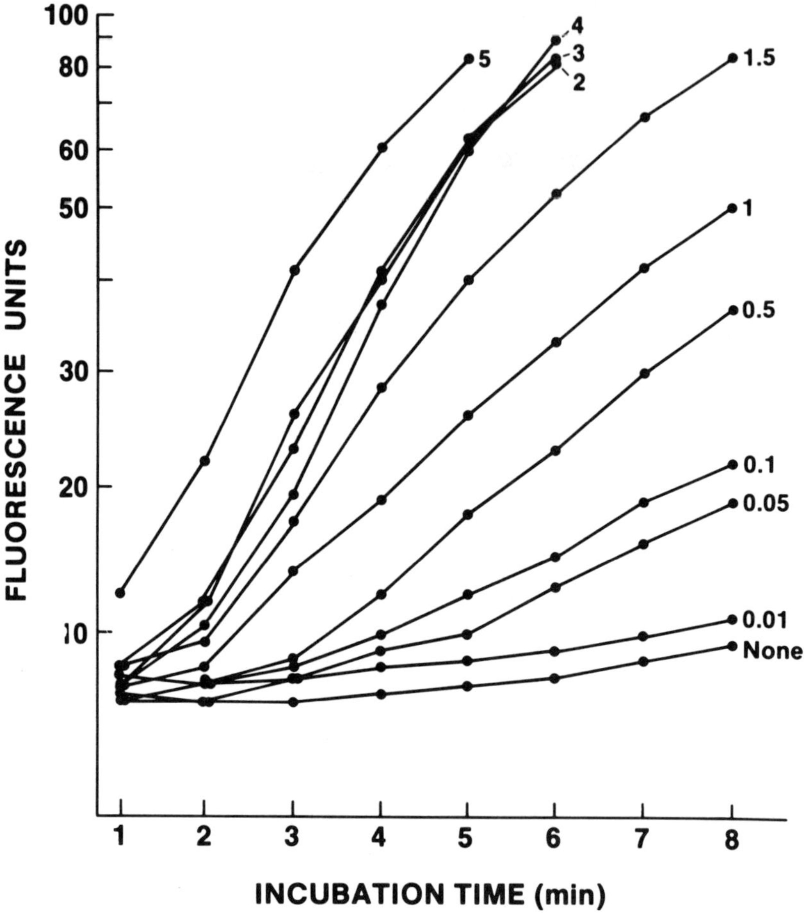

Figure 3: SMw32-P proteinase activity in the presence of various concentrations of glutathione (0.01-5.0 mM). Assays were done without pre-incubation of the reactants. Activity was measured with the substrate, CBZ-arg-arg-AFC. (reprinted with permission)

activation is attained. The data obtained at all GSH concentrations tested fall on the same monoexponential curve, indicating that enzyme activation kinetics are first order in GSH concentration. The second-order rate constant describing enzyme activation was determined to be 12.7 $M^{-1}min^{-1}$.

We had originally expected that activation of the proteinase might be subject to the ratios of oxidized (GSSG) to reduced glutathione. However, other experiments showed that half-maximal activation of the proteinase was

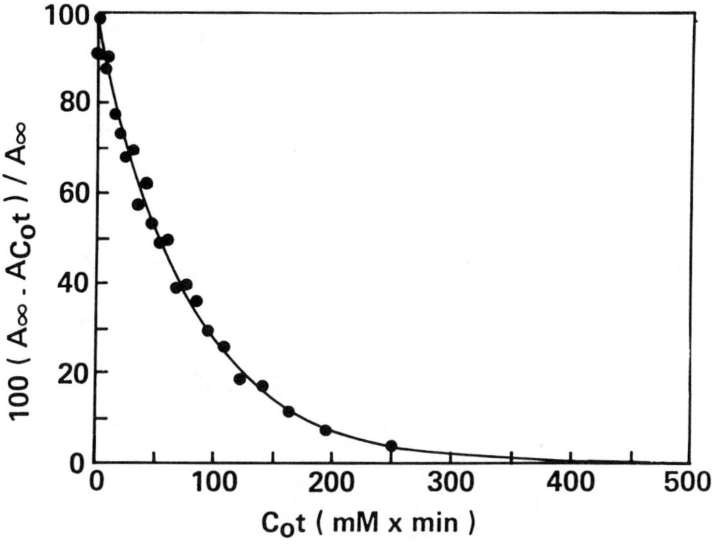

Time x Concentration Dependence of 32 kDa Activation

Figure 4: Reduced glutathione activation of the SMw32-P proteinase-Cot curve. Unactivated enzyme (10 μl/ml) was incubated in 2.5, 5.0 and 10 mM reduced GSH in 0.1 M citrate buffer (pH 5.5) and 1 mM EDTA. Aliquots were withdrawn at various times for assay with CBZ-arg-arg-AFC. The data were fitted by nonlinear least-squared analysis to the equation: $100(A - A_{Cot})/A = \exp(-kC_ot)$ (reprinted with permission)

at a ratio of GSH/GSSG of 1.0 (while physiological ratios are in the order of 100-300).

The actual residence time the proteinase is in the worm gut is limited since the parasite digestive tract contents are exchanged every 3-4 hours. Based on the data in Figure 6, the glutathione concentration we found in serum alone (24 μM) would allow only a limited activation of proteinase in this 4-hour period, at which time another "feeding cycle" would begin. In contrast, lysis of host red blood cells in the worm gut results in an increase in glutathione concentration to about 4 mM (29). At this GSH concentration, half-maximal activation of the proteinase occurs in 16 minutes.

Regulation of proteinase activation by thiols relies not on a chemical event but on the biological restraints (feeding cycle time) of the time required for activation. We conclude that the glutathione concentration in serum alone is insufficient to activate the proteinase within the feeding cycle time (3-4 hours). Red blood cell lysis in the parasite gut, however, raises the glutathione level more than 30-fold (from 0.13 to 4.2 mM, ref. 29) to concentrations

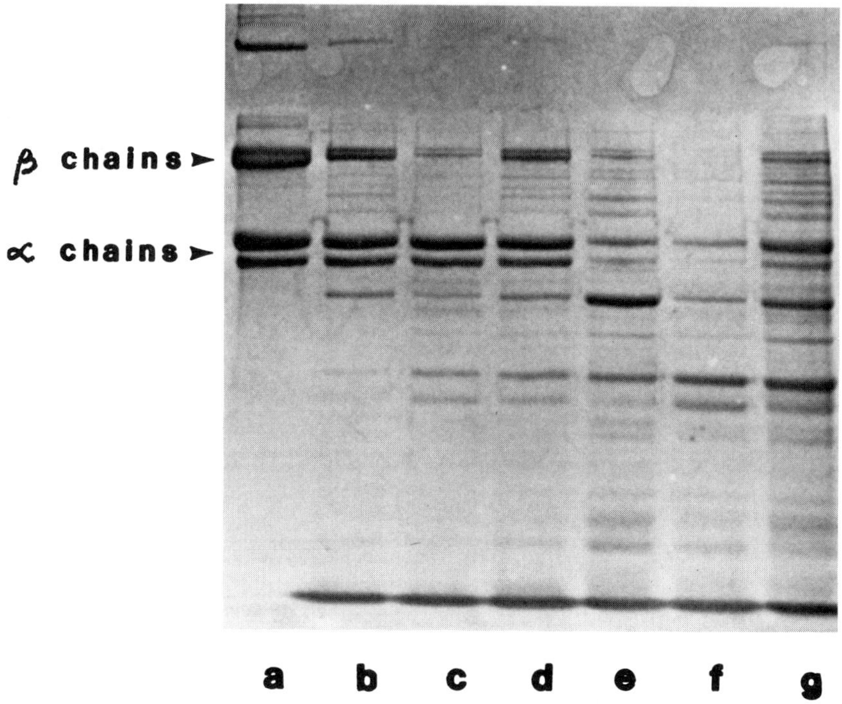

Figure 5: SDS-polyacrylamide gel electrophoresis pattern of collagenolysis by three species of *Strondyloides*: supernatant fractions were prepared from L_3 larvae of each species (2.5×10^6/cc 0.2 M Na citrate, pH 4.0). PBS washings were collected and concentrated. For collagenase assay, 50 μl (4 μg protein) of each enzyme were incubated in the standard assay mixture for 16 hr at 37°C. Samples were analyzed by 7.5% SDS-PAGE. a, guinea pig collagen (α_1, α_2, and β chains) control (no enzyme); b,c,d, collagenolytic products after incubation with supernatants from *S. stercoralis*, *S. ransomi*, and *S. ratti*, respectively; e,f,g, collagenolytic products after incubation with PBS washings from *S. stercoralis, S. ransomi*, and *S. ratti*, respectively. (reprinted with permission)

yielding significant activity in a short time. Therefore, the host blood ingested by *Schistosoma mansoni* adult worms contains not only the proteinase substrate, hemoglobin, but also the cofactor (glutathione) in concentrations sufficient to activate the digestive process.

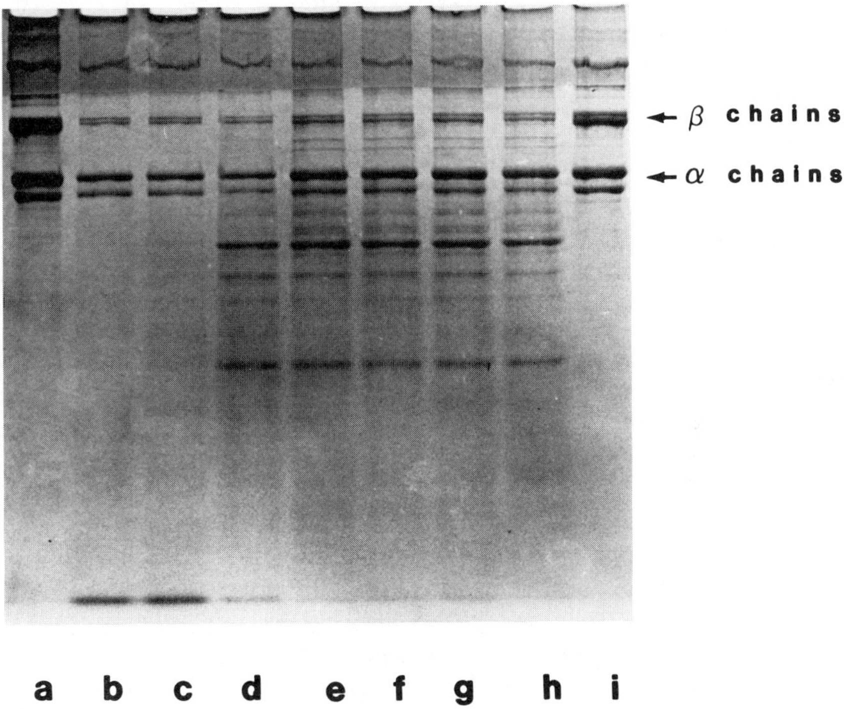

Figure 6: Effect of proteinase inhibitors on *Strongyloides ransomi* collagenolytic activity. Partially purified collagenolytic activity was assayed in the presence of thiols and various inhibitors. Samples were incubated for 16 hr at 37°C and then analyzed by 7.5% SDS-polyacrylamide gel electrophoresis. a, collagen control (no enzyme); collagenolysis in the presence of: b, no inhibitor; c, 10 mM EDTA; d, pepstatin (100 μg/ml); e, leupeptin (4 μg/ml); f, 2 mM iodoacetic acid; g, 10 mM N-ethylmaleimide; h, 10^{-5}M Ep459; i, pepstatin + Ep459. (reprinted with permission)

PENETRATION OF HELMINTHS THROUGH CONNECTIVE TISSUES

A wide variety of parasites are able to penetrate the connective tissues of their host organisms. Well-known examples are helminth cercariae and nematode L_3 larvae. The mechanisms by which a variety of helminths penetrate through skin have been reviewed by Stirewalt (31) and Matthews (32). Among enzymatic means, the collagenases have been commonly suggested as mediating connective tissue penetration. However, in the case of the helminth parasite, *Schistosoma mansoni*, it appears clear that penetration of skin can be accomplished without extensive digestion of the major connective tissue protein, collagen. In a series of studies started in 1972, we demonstrated, by testing connective tissue protein substrates *in vitro*, that the major proteinase involved in skin penetration is a serine proteinase (33-35). Its activity appears to be directed primarily against the protein backbone of the proteoglycan matrix of skin (36). In similar studies, McKerrow and his collaborators have shown that these cercarial proteinases are also active against elastin (37). Thus, while serine proteinases appear to be fully capable of connective tissue protein degradation, other proteinases able to degrade collagen have also been shown to mediate penetration through connective tissues.

Enzymes able to digest collagen were first described in anaerobic bacteria, which often result in extensive tissue digestion. Several of these enzymes were quite specific for collagen and had little activity on other protein substrates. In the early 1960s, Jerome Gross and his collaborators described the presence of such specific collagenases in a variety of eukaryotic, primarily vertebrate, tissues. The first reported vertebrate collagenase was from resorbing tadpole tissues (38). The vertebrate collagenases are remarkably alike in their enzymatic properties, although they differ in physical properties, such as molecular weight. These enzymes belong to the class of metalloendoproteinases and are able to cleave the triple helical portions of the collagen molecule at only one locus — in an area approximately 75% from the N-terminus of the triple helix. Moreover, as demonstrated in several cases, these enzymes are unable to hydrolyze either denatured collagen (gelatin) or other proteins such as casein or hemoglobin.

A second class of collagenolytic enzymes was described initially by Eisen and his collaborators (39,40), from the hepatopancreas of the fiddler crab, *Uca pugilator*. These collagenases are less specific than the vertebrate enzymes and are able to hydrolyze the low molecular weight substrates normally used to assess tryptic or chymotryptic enzymes. Inhibitor studies have revealed that, like trypsin and chymotrypsin, the invertebrate collagenases belong to the class of serine endoproteinases. In contrast to the vertebrate collagenases, these invertebrate enzymes make multiple cleavages in the triple helical portion of the collagen molecule. This activity differs from the vertebrate trypsins and chymotrypsins which hydrolyze the collagen triple helix only at very high concentrations.

The capacity of cysteine endopeptidases to hydrolyze collagen has not yet been clearly delineated. Part of the difficulty is due to the nature under which assays for these enzymes are done. The mammalian cysteine proteinases function at acidic pH, a condition in which triple-helical collagen may undergo denaturation to random-coil gelatin. Cysteine proteinases, such as cathepsin B, can remove the terminal telopeptide regions of collagen (generally considered not to be part of the collagen triple helix), which leads to a depolymerization of cross-linked collagen. Whether or not such enzymes cleave the triple-helical region of the native structure is less clear. In cases where there is cleavage of the triple helix, discrete cleavage products are not generated as they are with the mammalian collagenases (41).

Nevertheless, elevated levels of cysteine proteinases have been implicated in physiological situations where connective tissue destruction occurs. For example, Poole and his collaborators (42) have reported elevated levels of cathepsin B-like enzymes in invasive human tumors. Similarly, Baici *et al.* (43) suggested that tumors promoted the synthesis and secretion of cysteine proteinases in the host, to allow invasion of surrounding tissues. Evidence for an extracellular role of cysteine proteinases, rather than being solely lysosomal, is increasing. In an early study, Vaes and his collaborators (44) suggested that cysteine proteinases might play an extracellular role in bone resorption. More recently, Gal and Gottesman (45) have shown that the major excreted protein of retrovirus-transformed mouse cells is a glycoprotein with properties of the cysteine proteinase, cathepsin L. Studies in our laboratory, with the invasive nematode, *Strongyloides ransomi*, have revealed a parasite-produced cysteine proteinase that has endopeptidase activity against collagen (46). This activity results in multiple, discrete collagen cleavage products.

Thus, although acidic thiol proteinases are commonly thought of as lysosomal enzymes, they may play both extracellular and intracellular roles in protein degradation.

THE COLLAGENOLYTIC ACTIVITY OF *STRONGYLOIDES* SPECIES

The nematode parasite, *Strongyloides* spp., infects its definitive hosts by penetration through the skin. In the case of *Strongyloides* the most relevant studies are those of Lewert and his colleagues. In their work (47,48) using both histochemical and biochemical techniques, they determined that *Strongyloides ratti* altered the epidermal-dermal basement membrane, and that they contained proteolytic enzymes as measured by hydrolysis of Azocoll, an azo-dye coupled hide powder. Matthews also used this substrate to determine that *Strongyloides fulleborni* produced azocollytic proteinases while *Ancylostoma tubaeforme* and *Necator americanus*, which also infect by penetration, do not seem to contain such enzymes (49).

Evidence that *Strongyloides* produce proteolytic enzymes which may be important in infection seems fairly convincing; however, the nature of this enzyme(s) is less clear. Lewert and Lee (48) indicated that the proteinase from

Strongyloides ratti was most active against Azocoll and azocartilage at pH 7.0. The activity was not inhibited by trypsin inhibitors or compounds which alkylate SH groups (N-ethylmaleimide and iodoacetate). On the other hand, the activity was stimulated by some divalent cations (Mg^{++} and Ca^{++}), inhibited by others (Cu^{++}, Co^{++}, Hg^{++}) or by chelating agents, suggesting that the activity may be due to a metalloproteinase.

Our initial studies substantiate those of Lewert and his colleagues (45). We have demonstrated the presence of two groups of proteolytic enzymes by extraction of larvae at either acidic or neutral pH. However, examination of these extracts for collagenolytic activity revealed no activity at neutral pH. This observation is consistent with those of Wertheim *et al.* (50). Instead, collagenase activity was found in the acidic extract only (46). Extracts from the L3 larvae of three different *Strongyloides* species, *S. stercoralis* (human host), *S. ratti* (rat host), and *S. ransomi* (pig host) displayed similar collagenase activity (Figure 5). The discrete banding pattern below the chains indicates that these enzymes cleave both the N-terminal telopeptide, crosslink-containing region (converting β chains to α chains) and the triple-helical region of collagen. The former activity is also seen with vertebrate cathepsin B (41) but not the latter pattern. Activity of the parasite collagenase is maximal at pH 4.0-4.5 with essentially no activity observed above pH 7.0. Collagenase activity was measured in the presence of various proteinase inhibitors (Figure 6). The inhibitors pepstatin (aspartate proteinases) and EDTA (metalloproteinases) had no effect on collagenase activity; however, inhibitors of cysteine porteinases, such as iodoacetate, leupeptin and Ep459 completely abolished collagenolysis.

Activity was also found in the supernatant of frozen and thawed larvae, suggesting that these larvae secrete enzymes during penetration. The mechanism of penetration must be investigated further. Since these enzymes are dependent on thiols and acidic pH for activity, the appropriate environment to activate the enzymes during penetration must be demonstrated. Nevertheless, these studies clearly describe proteinase activity in parasitic organisms able to digest the collagen triple helix. These observations lend support to a new extracellular role for such enzymes.

ACKNOWLEDGEMENTS

I want to express my deep and sincere thanks to Mahlon B. Hoagland for introducing me to the excitement of scientific research. He provided the impetus for developing a career as an independent investigator and the opportunity to learn to do good experiments. I also thank Oscar Scornik for his interest and concern during my graduate student days.

I also thank my colleagues Cynthia Chappell and Ajay Rege for their research results. This work was supported by grants from the National Institutes of Health, the National Science Foundation, and the U.S. Department of Agriculture.

REFERENCES

1. Timms, A. R., and Bueding, E. (1959). Studies of a proteolytic enzyme from *Schistosoma mansoni*. *Brit. J. Pharm.*, **14**:68-73.
2. Grant, C. T., and Senft, A. W. (1971). *Schistosome proteolytic enzyme, Comp. Biochem. Physiol.*, **38B**:663-678.
3. Sauer, M. C. V., and Senft, A. W. (1972). Properties of a proteolytic enzyme from *Schistosoma mansoni*. *Comp. Biochem. Physiol.*, **42B**:205-220.
4. Dresden, M. H., and Deelder, A. M. (1979). *Schistosoma mansoni*: Thiol proteinase properties of adult worm "hemoglobinase." *Exptl. Parasitol.*, **48**:190-197.
5. Chappell, C. L., and Dresden, M. H. (1986). *Schistosoma mansoni*: Proteinase activity of "hemoglobinase" from the digestive tract of adult worms. *Exptl. Parasitol.*, **61**:160-167.
6. Chappell, C. L., and Dresden (1986). Characterization of a cysteinyl proteinase from the human parasite *Schistosoma mansoni*. **In**: *Cysteine Proteinases and Their Inhibitors*, V. Turk, (Ed.), Walter de Gruyter, Berlin, pp. 199-208.
7. Lindquist, R. N., Senft, A., Petitt, M., and McKerrow, J. H. (1986). *Schistosoma mansoni*: Purification and characterization of the major acidic proteinase from adult worms. *Exptl. Parasitol.*, **61**:398-404.
8. Locatelli, A., and Paoletti, C. (1969). *Fasciola hepatica* L. homogenate proteolytic activity: Action against different substrates. *Archivio Veterinario Italiano*, **20**:97-100.
9. Rupova, L., and Keilova, H. (1979). Isolation and some properties of an acid protease from *Fasciola hepatica*. *Z. Parasitenkunde*, **61**:83-91.
10. Aoki, T., and Oya, H. (1978). Hemoglobin-specific acid proteases in *Schistosoma mansoni*, *Angiostrongylus cantonensis*, and *Fasciola hepatica*: Multiple forms of the enzymes on polyacrylamide gels coated with hemoglobin. *Japanese J. Parasitol.*, **27**:(Suppl), 101.
11. Simpkin, K. G., Chapman, C. R., and Coles, G. C. (1980). *Fasciola hepatica*: A proteolytic digestive enzyme. *Exptl. Parasitol.*, **49**:281-287.
12. Sato, K., Takahashi, J., and Sawada, T. (1976). Studies on the globinolytic enzymes of *Dirofilaria immitis, Schistosoma japonicum*, and *Clonorchis sinensis*. *Japanese J. Parasitol.*, **25**:8-9.
13. Maki, J., Furuhashi, A., and Yanagisawa, T. (1982). The activity of acid proteases hydrolysing haemoglobin in parasitic helminths with special reference to interspecific and intraspecific distribution. *Parasitology*, **84**:137-147.
14. Lumley, A. M., and Lee, D. L. (1980). Changes in nematodes during development of immunity to these nematodes in their hosts: Proteolytic enzymes. **In**: *The Host-Invader Interplay*, H. van den Bossche, (Ed.), Elsevier-North Holland, Amsterdam, pp. 443-455.
15. Oya, Y., and Noguchi, I. (1977). Some properties of hemoglobin protease from *Ancylostoma caninum*. *Japanese J. Parasitol.*, **26**:307-313.
16. Keene, W. E., Petitt, M. G., Allen, S., and McKerrow, J. H. (1986). The major neutral proteinase of *Entamoeba histolytica*. *J. Exp. Med.*, **163**:536-549.
17. Lushbaugh, W. B., Hoffbauer, A. F., and Pittman, F. E. (1984). Proteinase activities of *Entamoeba histolytica* cytotoxin. *Gastroenterol.*, **87**:17-27.
18. Lushbaugh, W. B., Hoffbauer, A. F., and Pittman, F. E. (1985). *Entamoeba histolytica*: Purification of cathepsin B. *Exptl. Parasitol.*, **59**:328-336.
19. Scholze, H., and Werries, E. (1984) A weakly acidic protease has a powerful proteolytic activity in *Entamoeba histolytica*. *Molec. Biochem. Parasitol.*, **11**:293-300.

20. Scholze, H., and Werries, E. (1986). Cysteine proteinase of *Entamoeba histolytica*. I. Partial purification and action on different enzymes. *Molec. Biochem. Parasitol.*, **18**:103-112.
21. McLaughlin, J., and Muller, M. (1979). Purification and characterization of a low molecular thiol proteinase from the flagellate protozoan. *Tritrichomonas foetus*. *J. Biol. Chem.*, **254**:1526-1533.
22. Lonsdale-Eccles, J. D., and Mpimbaza, G. W. N. (1986). Thiol-dependent proteases of African trypanosomes. Analysis by electrophoresis in sodium dodecyl sulphate/polyacrylamide gels co-polymerized with fibrinogen. *Eur. J. Biochem.*, **155**:469-473.
23. Bolla, R., and Weinstein, P. P. (1980). Acid Protease activity during development and aging of *Nippostrongylus braziliensis*. *Comp. Biochem. Physiol.*, **66B**:475-481.
24. Bogitsh, B. J. (1978). *Schistosoma mansoni*: Uptake of exogenous hemeproteins by schistosomula grown *in vitro*. *Exptl. Parasitol.*, **45**:247-254.
25. Lawrence, J. D. (1973). The ingestion of red blood cells by *Schistosoma mansoni*. *J. Parasitol.*, **59**:60-63.
26. Chappell, C. L., and Dresden, M. H. (1987). Purification of cysteine proteinases from adult *Schistosoma mansoni*. *Arch. Biochem. Biophys.*, **256**:560-568.
27. Hanada, K., Tamai, M., Morimoto, S., Adachi, T., Oguma, K. Ohmura, S., and Ohzeki M. (1980). A specific thiol protease inhibitor, E-64, and its derivatives. *Peptide Chem.*, **1979**:31-39.
28. Barrett, A. J., Kembhavi, Brown, M. A., Kirschke, H., Knight, C. G., Tamai, M., and Hanada, K. (1982). L-trans-Epoxysuccinyl-leucylamido (4-guanidino) butane (E-64) and its analogs as inhibitors of cysteine proteinases including cathepsins B, H and L. *Biochem. J.*, **201**:189-198.
29. Chappell, C. L., Dresden, M. H., Walters, D. W. (1987). Glutathione activation of a cysteine proteinase from *Schistosoma mansoni*. *Biochim. Biophys. Acta* **913**:335-341.
30. Meister, A. (1984). New aspects of glutathione biochemistry and transport: selective alteration of glutathione metabolism. *Fed. Proc.*, **43**:3031-3042.
31. Stirewalt, M. A. (1972). Skin penetration mechanisms of helminths, **In**: *Biology of Parasites*, E. J. L. Soulsby, Ed., Academic Press, New York, pp. 41-59.
32. Matthews, B. E. (1977). The passage of larval helminths through tissue barriers, **In**: *Parasite Invasion*, Symposia of the British Society for Parasitology, vol. 15, A. E. R. Taylor, and R. Muller, Eds., Blackwell Scientific Publications, Oxford, pp. 93-119.
33. Dresden, M. H., and Asch, H. L. (1972). Proteolytic enzymes in extracts of *Schistosoma mansoni* cercariae. *Biochem. Biophys. Acta*, **289**:378-384.
34. Dresden, M. H., Lewis, J. C., and Krisko, I. (1977). Proteolytic action of *Schistosoma mansoni* cercarial proteases on keratin and basement membrane proteins. *J. Parasitol.*, **63**:941-943.
35. Landsperger, W. J., Stirewalt, M. A., Dresden, M. H. (1982). Purification and properties of a proteolytic enzyme from the cercariae of the human trematode parasite *Schistosom mansoni*. *Biochem. J.*, **201**:137-144.
36. Dresden, M. H. (1982). Proteolytic enzymes of *Schistosoma mansoni*. *Acta Leidensia*, **49**:81-99.
37. McKerrow, J. H., Jones, P. Sage, H., and Pino-Heiss S. (1985). Proteinases from invasive larvae of the trematode parasite *Schistosoma mansoni* degrade connective-tissue and basement-membrane macromolecules. *Biochem. J.*, **231**:47-51.
38. Gross, J., and Lapiere, C. M. (1962). Collagenolytic activity in amphibian tissues: a tissue culture assay. *Proc. Natl. Acad. Sci. USA*, **48**:1014-1022.
39. Eisen, A. Z., and Jeffrey, J. J. (1969). An extractable collagenase from crustacean hepatopancreas. *Biochem. Biophys. Acta*, **191**:517-526.
40. Grant, C. A., Henderson, K. O., Eisen, A. Z., Bradshaw, R. A. (1980). Amino acid sequence of a collagenolytic protease from the hepatopancreas of the fiddler crab. *Uca pugilator*, *Biochem.*, **19**:4653-4659.

41. Burleigh, M. C. (1977). Degradation of collagen by non-specific proteinases. **In**: *Proteinases in Mammalian Cells and Tissues.* A. J. Barrett, Ed., North-Holland, Amsterdam, pp. 285-309.
42. Recklies, A. D., Mort, J. S., and Poole, A. R. (1982). Secretion of a thiol proteinase from mouse mammary carcinomas and non-malignant human breast tissue. *Cancer Res.*, **42**:1026-1032.
43. Baici, A., Gyger-Marazzi, M., and Strauli, P. (9184). Extracellular cysteine proteinase and collagenase activities as a consequence of tumor-host interaction in the rabbit V2 carcinoma. Invasion Metastasis, **4**:13-27.
44. Delaisse, J.-M., Eeckhort, Y, and Vaes, G. (1980). inhibition of bone resorption in culture by inhibitors of thiol proteinases. *Biochem. J.*, **192**:365-368.
45. Gal, S., and Gottesman, M. M. (1986). The major excreted protein (MEP) of transformed mouse cells and cathepsin L have similar protease specificities. *Biochem. Biophys. Res. Comm.*, **139**:156-162.
46. Rege, A. A., Dresden, M. H. (1987). *Strongyloides* spp.: Demonstration and partial characterization of acidic collagenolytic activity from infective larvae. *Exptl. Parasitol.*, **64**:275-280.
47. Lewert, R. M., and Lee, C. L. (1954). Studies on the passage of helminth larvae through host tissues. I. Histochemical studies on the extracellular changes caused by penetrating larvae. II. Enzymatic activity of larvae *in vitro* and *in vivo*. *J. Infect. Dis.*, **95**:13-51.
48. Lewert, P. M., and Lee, C. L. (1956). Quantitative studies on the collagenase-like enzymes of cercariae of *Schistosoma mansoni* and the larvae of *Strongyloides ratti*. *J. Infect. Dis.*, **99**:1-14.
49. Matthews, B. E. (1975). Mechanism of skin penetration by *Ancylostoma tubaeforme* larvae. *Parasitol.*, **70**:25-38.
50. Wertheim, G., Lustigman, S., Silberman, H., and Shoshan, S. (1983). Demonstration of collagenase activity in adult *Strongyloides ratti* (Nematoda: Strongyloididen) and its absence in infective larvae. *J. Helminthol.*, **57:241-246.**

8

SELECTIVE INHIBITION OF ISOPENTENYL PYROPHOSPHATE AND SQUALENE SYNTHESIS

Thomas S. Parker[1]

The Rockefeller University
1230 York Avenue
New York, NY 10021

Barry Gotlinsky[2], Robert Engel, and Burton E. Tropp[3]

Doctoral Programs in Chemistry and Biochemistry
The City University of New York
Queens College
Flushing, NY 11367

[1] Current address: Department of Biochemistry
Cornell Medical Center
429 E. 71st Street
New York, NY 10021

[2] Current address: PAL Industries
73 Bloomingdale Road
Hicksville, NY 11801

[3] Postdoctural fellow with Mahlon B. Hoagland
Harvard Medical School

SUMMARY

When added to a cell-free multienzyme system prepared from rat liver (S_{10}), 5-carboxy-4-hydroxy-4-methylpentyl-1-phosphonic acid (5-homomethylenemevalonophosphonate, MMP), a phosphomevalonic acid analog, was found to be a potent inhibitor of isopentenyl pyrophosphate (IPP) synthesis. The analog did not inhibit cholesterol synthesis by intact rat hepatocytes in monolayer culture presumably because highly charged substances in general cannot cross cytoplasmic membranes in the absence of a suitable carrier system. This problem was overcome by coupling MMP through its carboxyl group to the 24-hydroxyl of 3α,7α,12α -trihydroxy-5β-cholan-24-ol (reduced cholic acid) to form a conjugate, structurally related to taurocholic acid, for which there is a natural transport system in hepatocytes. When presented to hepatocytes as the conjugate, MMP inhibited the conversion of mevalonic acid to cholesterol and when administered intravenously to rats, conjugated (but not unconjugated) [^{14}C]-MMP was concentrated within the liver and intestine and excreted in bile.

Phosphonate analogs of natural organic phosphates often act as metabolic inhibitors (1). The replacement of the -O-P bond of the phosphate by the $-CH_2-P$ structure of the phosphonate linkage makes these analogs resistant to hydrolysis by phosphatase. Because of their structural relationship to the natural phosphates, phosphonate analogs have the potential to serve as both chemotherapeutic agents and metabolic probes. However, there is one very serious limitation to the use of phosphonate analogs for these purposes; cell membranes are generally impermeable to organic phosphonates.

It has been suggested that selective inhibition of the enzymes which convert isopentenyl pyrophosphate (IPP) to squalene would preserve the supply of mevalonic acid for the production of essential metabolites while inhibiting cholesterol synthesis (2). However, the design of the selective inhibitors of enzymes that occur in this pathway segment has proven difficult because the natural substrates are either phosphate or diphosphate esters. Substrate analogs lacking the polar phosphate ester function are poor inhibitors, while those having it are susceptible to inactivation by hydrolytic enzymes and are not taken up readily by intact cells. Corey and Volante (3), Popjak *et al.* (4), and Parker *et al.* (5) have synthesized phosphonate analogs that inhibit the conversion of mevalonic acid to squalene in cell-free systems. In a previous report, we demonstrated that 5-carboxy-4-hydroxy-4-methylpentyl-1-phosphonic acid (5-homomethylenemevalonophosphonate, MMP), an isosteric analog of 5-phosphomevalonic acid, is an inhibitor of 5-phosphomevalonate phosphokinase (4). We have now labeled MMP with ^{14}C and used it to test a delivery system designed to promote uptake of phosphonate-bearing inhibitors by cells having bile salt transport systems. MMP was coupled to a bile salt derivative through an ester link between its carboxyl group and the 24-hydroxyl of 3α,7α,12α,-trihydroxy-5β-cholan-24-ol (reduced cholic acid). The resulting conjugate mimicked the structure of taurocholic acid, was cleared from the circulation by the liver when injected into rats, and inhibited cholesterol synthesis from mevalonic acid in cultured rat hepatocytes. These results suggest that the bile transport system may be exploited for the illicit transport of other phosphonate analogs.

MATERIALS AND METHODS

General

All chemicals were of commercial reagent quality and were used without further purification with the following exceptions: pyridine and diisopropylamine were distilled and stored over KOH pellets; hexane and benzene were dried over sodium metal; tetrahydrofuran was distilled over lithium aluminum hydride and stored over molecular sieves; and naphthalene was recrystallized from 2-propanol immediately before use.

(*RS*)-[1-^{14}C]-Mevalonic acid lactone (sp. act. 6.0 Ci/mol) and (*RS*)-[2-^{14}C]mevalonic acid lactone (sp. act. 2.5 Ci/mol) were obtained from Amersham, Arlington Heights, IL. (*RS*)-[5-^{3}H]-mevalonic acid (sp. act. 7.8 Ci/mol), sodium [2-^{14}C]-acetate (sp. act. 10 Ci/mol) and [24-^{14}C]-taurocholic acid (sp. act. 50 Ci/mol) were obtained from New England Nuclear, Boston, MA. The synthetic procedures used to prepare diethyl 4-oxopentyl-1-phosphonate (I), diethyl 5-carboxy-4-methylpentyl-1-phosphonate, (II) 5-carboxy-4-hydroxy-4-methylpentyl-1-phosphonic acid (MMP) (III), 3α,7α,12α,-trihydroxy-5-β-cholan-24-ol [systematic name: 17β-(1-methyl-4-hydroxybutyl) etiocholane-3α,7α,12α-triol); trivial name cholyl alcohol] (IV) diethyl-5-carbo (3α,7α,12α-trihydroxy-5β-cholan-24-hydroxyl) oxy-4-hydroxyl-4-methylpentyl-1-phosphonate (cholyl diethyl MMP) (V), and 5-carbo(3α,7α,12α-trihydroxy-5β-cholan-24-hydroxyl) oxy-4-hydroxy-4-methylpentyl- 1-phosphonic acid (cholyl-MMP) (VI) are outlined in Figure 1. These methods have been described in detail elsewhere (6,7). [5-^{14}C]-5-carbo(3α,7α,12α-trihydroxy-5β-cholan-24-hydroxyl) oxy-4-hydroxy-4-methylpentyl-1-phosphonic acid (cholyl-[5-^{14}C]-MMP) (VII) was prepared by the previously described route (6,7) using 25 mCi of sodium [2-^{14}C]-acetate (sp. act. 10 Ci/mol). An overall yield of 28% of labeled material was obtained.

Fetal calf serum (FCS), arginine-free Eagle's basal medium as modified by Dulbecco (DME), penicillin G (50,000 units/ml), Hank's buffered saline, phosphate-buffered saline (PBS), and trypsin in EDTA (0.02%) were obtained from Grand Island Biological Co., Santa Clara, CA. FCS was inactivated at 56°C for 3 min and stored at -20°C. Media were supplemented before use with 0.4 mM ornithine. Collagenase type I, porcine insulin, bovine serum albumin (essentially fatty acid free), and ornithine were obtained from Sigma Chemical Co., St. Louis, MO. Plastic petri dishes, 60mm in diameter, were obtained from Lux, Inc.

Animals

Male Sprague-Dawley rats used for preparations of 10,000 × g supernatants of liver homogenates (S_{10}) were from the breeding colony of Dr. S. Roberts, Dept. of Biological Chemistry, School of Medicine, University of California, Los Angeles, CA. All other studies were carried out in pathogen-free female rats obtained from the Charles River Breeding Laboratories, Wilmington, MA. All rats received Purina chow and water *ad libitum* and were adapted for two weeks or longer to a normal light cycle (light 6 am to 6 pm).

Rat Liver S_{10} Preparation

Rats (90-200 g) were killed between 8 and 10 am by spinal dislocation for preparation of the 10,000 × g × 20 min supernatants as described by Popjak (8) except that the homogenization buffer contained 200 mM potassium phosphate (pH 7.5 at 4°C), 60 mM nicotinamide, 100 mM dithiothreitol, and 10 mM $MgCl_2$. S_{10} preparations were used immediately to minimize time- and

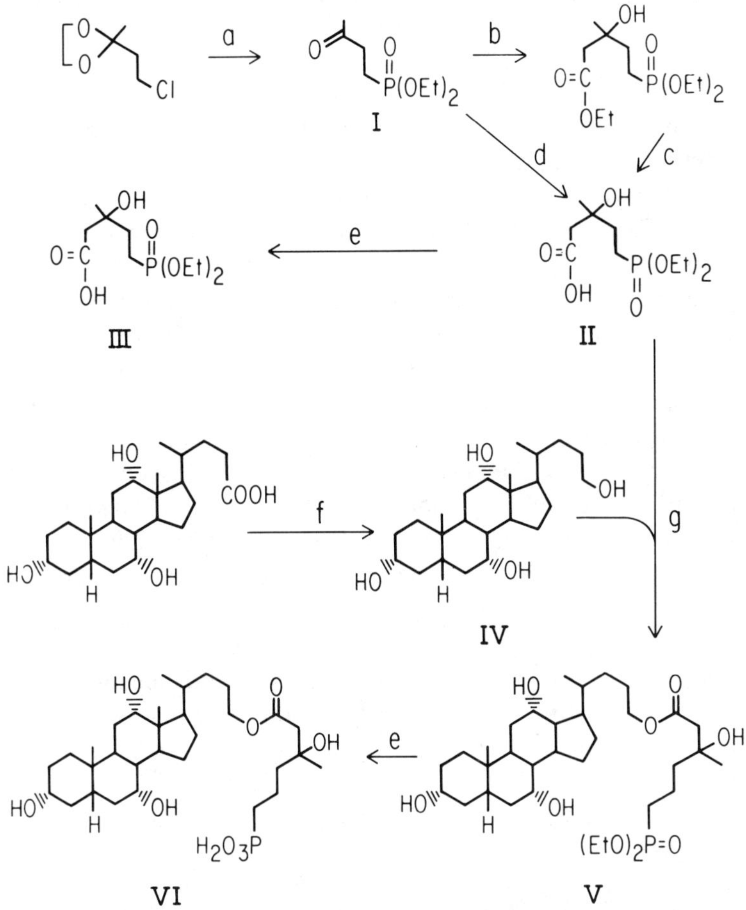

Figure 1: Synthesis of MMP and its coupling to cholyl alcohol. Reagents indicated as follows. a: (1) triethyl phosphite. (2) aqueous acid. b: lithium diisopropylamide with ethyl acetate. c: methanolic potassium hydroxide. d: (1) lithium, naphthalene and sodium acetate, (2) water. e: (1) chlorotrimethylsilane, (2) water. f: lithium aluminum hydride. g: triisopropylbenzene sulfonyl chloride. Numbered compounds are: I, diethyl 4-oxopentyl-1-phosphonate; II, diethyl 5-carboxy-4-hydroxy-4-methylpentyl-1-phosphonate or diethyl 5-homomethlenemevalonophosphonate (diethyl MMP); III, 5-carboxy-4-hydroxy-4-methylpentyl-1-phosphonate or 5-homomethylenemevalono-1-phosphonate (MMP); IV, 3α,7α,12α-trihydroxy- 5-β-cholan-24-01 (cholyl alcohol); V, diethyl-5-carbo(3α,7α,12α-trihydroxy-5β-cholan-24-hydroxyl)oxy-4-hydroxy-4-methylpentyl-1-phosphonate (cholyl-diethyl MMP); VI, 5-carbo(3α,7α,12α-trihydroxy-5-β-cholan-24-hydroxyl)oxy-4-hydroxy-4-methylpentyl-1-phosphonate (cholyl-MMP).

temperature-dependent changes in enzyme activities and cofactor requirements (9,10).

Rat Liver S_{10} Incubations

Incubations were performed in 16 × 60 mm screw cap tubes to facilitate the collection of $^{14}CO_2$ and subsequent saponification and extraction of the neutral lipid fractions. Each tube containing 200 μL of fresh S_{10} and either 300 μL of water or a solution of test compound was sealed with a rubber-serum cap. The caps held a center well cup (Kontes) containing 400 μEq of hyamine hydroxide dispersed on a fluted 4 × 10 mm wick of Whatman paper for CO_2 collection (the methanol contained in preparations of hyamine must be evaporated from the cups to prevent inactivation of the S_{10} enzymes).

The tubes were evacuated and filled with N_2 three times through vents (18 gauge needles) to remove O_2 and were kept on ice until needed. Incubations were started after preincubation at 37°C for 1 min by injection of 500 μL of a mixture consisting of 300 μL of homogenization buffer and 50 μL each of 60 mM ATP, 60 mM glucose 6-phosphate (G6P), 20mM NADP, and 2.0 mM potassium mevalonate (50 nCi each of [1-^{14}C]- and [2-^{14}C]-mevalonate.

Incubations were stopped by injection of 1 M citric acid (1 mL) and $^{14}CO_2$ was collected at 50°C for 1 h. The $^{14}CO_2$ was measured by scintillation counting of the CO_2 collection cups in 10 mL of scintillation medium 2a70B (RPI Corp.) The hydrocarbon (mostly squalene) and nonpolar alcohol fractions were extracted and resolved by column chromatography on alumina as described by Popjak (8). After evaporation of solvent, the ^{14}C content of these fractions was measured by scintillation counting in 10 mL of Liquifluor (New England Nuclear).

Recoveries of $^{14}CO_2$ and [^{14}C]-squalene added to the incubation system were 95-100% and 90-95%, respectively. There was no detectable cross-contamination of $^{14}CO_2$ and [^{14}C]-squalene. The nonpolar alcohols (mostly sterol and free farnesol) accounted for less than 15% of the radioactivity in the neutral lipid fraction. The remaining 85% was isolated in the hydrocarbon fraction of which more than 85% was squalene.

Hepatocyte Monolayers

Hepatocytes were prepared from female rats (90-150 g) by the method of Berry and Friend (11) as modified by Edwards (12). Separated hepatocytes were washed three times by gentle suspension in DME containing 20% FCS and centrifugation at 300 RPM for 10 s (no brake). The cells, suspended in DME containing 20% FCS, insulin (10 μg/mL), streptomycin sulfate (100 μg/mL), and penicillin G (100 units/mL), were plated onto 60 mm plastic culture dishes and incubated at 37°C under an atmosphere of 95% air and 5% CO_2.

Rat hepatocytes were obtained in a yield of 0.5 to 4.5 × 10^8 cells per liver (from 90 -150 g rats; n = 12). In most preparations more than 90% of the cells excluded trypan blue (94±4%; n = 12); preparations were not used if less than 90% of the cells excluded trypan blue. Cell preparations obtained in this way attached to culture dishes within 5 h and gave monolayers containing 1-3 mg of protein per dish.

After 24 h in culture the medium was removed with any non-adherent cells and replaced with serum-free DME containing antibiotics and insulin and the monolayers were returned to the incubator. For studies of cholesterol synthesis the medium was replaced after a 3 h incubation with fresh serum-free medium containing either PBS or solutions of test compounds in PBS. The rate of cholesterol synthesis from mevalonic acid was measured by addition of (*RS*)-[5-^{3}H]mevalonic acid lactone (1 μCi in 200 μL) to give a final concentration of 1 μM, followed by incubation at 37°C under an atmosphere of 95% O_2 and 5% CO_2. Incubations were stopped by addition of 300 μL of 1 N KOH. The monolayers were dissolved at 70°C for 30 min and samples were taken for protein determination. Squalene, cholesterol, and lanosterol were added as carrier (1 mg each) and the mixture was saponified in 50% ethanol, 1 N KOH at 70°C for 1 h. The neutral lipid fraction was obtained by three extractions into light petroleum (bp 40-60°C), concentrated under a stream of nitrogen, and resolved into squalene, cholesterol, and methyl sterol fractions by thin-layer chromatography on silica gel G (Redi-coat G; Supelco) with toluene:ethyl acetate (7:3) as the solvent.

Whole Animal Studies

Female rats (200-250 g) were fitted with jugular vein catheters (PE 20) and bile duct catheters (PE 10) under light anesthesia with diethyl ether. The catheters were fixed in place with 5-0 silk and exteriorized through the surgical wounds. The wounds were closed in a single layer and the animals remained under anesthesia for the duration of the experiment (up to 120 min). Body temperatures of the animals kept under a heat lamp were monitored with a rectal probe and maintained between 37° and 39°.

Test compounds dissolved in 200 μL of saline were administered through the jugular vein catheter (dead volume less than 20 μL). complete delivery was assured by clearing the catheter with a 600 μL wash of saline. This procedure cleared 99.9% of the injected material from the catheter.

Table 1: Standard incubation conditions for testing inhibitors of isopentenyl pyrophosphate or squalene synthesis from mevalonic acid by the rat liver S_{10} multienzyme system.

Component	Amount
S_{10} Protein	5-6 mg/ml
Potassium phosphate buffer (pH7.5)	100 mM
Nicotinamide	30 mM
Dithiothreitol (DTT)	10 mM
Magnesium Chloride	5 mM
ATP	3 mM
Glucose 6-phosphate	3 mM
NADP	1 mM
RS-[1-^{14}C]-Mevalonic acid	50 nCi
RS-[2-^{14}C]-Mevalonic acid	50 nCi
Total *RS*-Mevalonic acid	100 uM
Atmosphere	100% N_2

Table 2: Effects of Changes in the Initial Concentrations of Cofactors and Intermediates on the Steady State rates of IPP and Squalene Synthesis from Mevalonic Acid.

Addition to the Complete System	Steady State Rate (Percent of Control)	
	IPP	Squalene
None	100	100
ATP (6 mM)	142	59
G6P + NADP (6 and 2 mM)	93	89
NADPH (1 mM)	98	59
DTT (20 mM)	93	89
NaF (10 mM)	29	61
DMAPP (200 μM)	94	207
GPP (200 μM)	70	120

The rates of IPP and squalene synthesis in the control incubations were 2.5 and 0.9 nCi/min, respectively. All steady state rates are calculated from 3 or more timed incubations taken during the linear portion of the reaction progress curve.

Bile was collected at 10 min intervals, the volumes were measured in calibrated microliter syringes (Hamilton Syringe Co.) and portions were dispersed into 10 mL Aquasol for scintillation counting. Bile flow averaged 3 mL and 2 mL in the first and second hours, respectively.

At the end of each experiment, the animals were exsanguinated by cardiac puncture. Blood and samples of various tissues were collected, weighed, and homogenized with an equal weight of saline in a Tissuemizer (Kontes). The amount of ^{14}C and ^{3}H labeled material in each tissue was determined by combustion of weighed and dried portions in a Packard Tri-Carb Sample

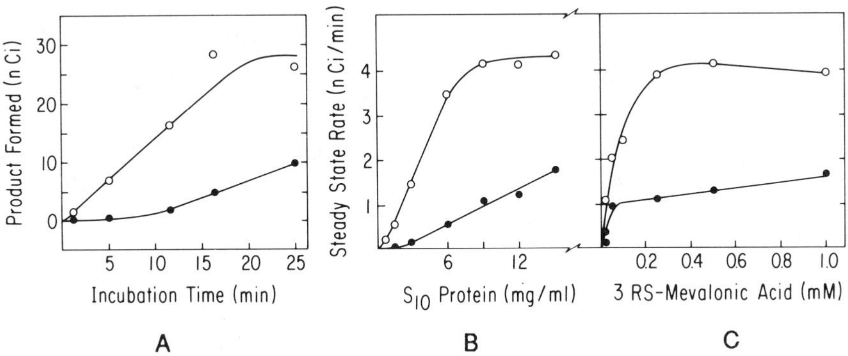

Figure 2: Kinetics of IPP (O) and squalene (●) synthesis by the rat liver S_{10} system. The kinetics of squalene synthesis from mevalonic acid by the rat liver S_{10} multienzyme system were characterized by dividing the reaction sequence into two pathway segments: the rate of the first three reactions converting mevalonate to IPP was measured as $^{14}CO_2$ production from [1-^{14}C]-mevalonate and the rate of squalene synthesis from IPP was measured as ^{14}C-squalene production from [2-^{14}C]-mevalonate. A) Time course of IPP (O) and squalene (●) synthesis by the standard system (see Table 1) showing the lag period, the steady state and plateau phases of the reactions; B) Relation of steady state rate and protein concentration; C) Steady state reaction velocity *vs.* the *RS*-mevalonic acid concentration.

Oxidizer Model 306, and collection and scintillation counting of $^{14}CO_2$ and ^3H_2O, respectively.

RESULTS

The 10,000 × g supernatant (S_{10}) of homogenized rat liver contains all of the enzymes needed to convert mevalonic acid to cholesterol. When supplemented as shown in Table 1 and incubated under an atmosphere of nitrogen to block squalene epoxidation, [1-^{14}C]-mevalonic acid is converted in three ATP-dependent steps to $^{14}CO_2$ and IPP, while [2-^{14}C]-mevalonic acid is converted to [^{14}C]-squalene. Under these conditions IPP synthesis, measured by $^{14}CO_2$ release and [^{14}C]-squalene synthesis were linear with time from 2-15 min and from 7-30 min, respectively as shown in panel A of Figure 2. The steady state reaction rates over each pathway segment were calculated from the linear portions of their respective progress curves. The relationships of the calculated steady state rate relative to the concentrations of S_{10} protein and mevalonic acid are shown in panels B and C. At all concentrations of mevalonic acid and protein studied, the synthesis of IPP was faster than that of squalene.

The steady state reaction rates of the two pathway segments were measured under a variety of conditions shown in Table 2. The rates of IPP and squalene synthesis could not be equalized by changing the cofactor concentrations. Doubling the ATP concentration stimulated the three ATP-dependent steps between mevalonic acid and IPP, but inhibited squalene synthesis. Sodium fluoride, reported to protect the intermediate prenyl diphosphates from hydrolysis by phosphatase (13), inhibited ATP-dependent steps of IPP synthesis and did not improve squalene yield. Doubling the capacity of the NADPH-generating system by adding an equivalent amount of NADPH had no effect on IPP synthesis but decreased squalene synthesis.

In contrast, addition of the intermediate substrates, 3,3-dimethylallyl- or geranyl pyrophosphate (DMAPP or GPP), did equalize the rates of IPP and squalene synthesis by specifically increasing synthesis of squalene but not that of IPP. This increase, which is a minimum estimate since isotopic dilution by the added unlabeled DMAPP was not taken into account, indicates that the overall rate-limiting step under these conditions occurs after IPP formation and probably at the reaction catalyzed by IPP isomerase. This fact makes it possible to estimate the rate of squalene synthesis from IPP (the second pathway segment) by measuring the rate of ^{14}C-squalene production from [2-^{14}C]-mevalonic acid. MMP inhibits the ability of the S_{10} system to synthesize both IPP and squalene. The concentrations of MMP required to give a 50% inhibition of IPP and squalene are 0.14 mM and 0.40 mM, respectively.

MMP was incubated with freshly isolated rat hepatocytes. At concentrations below those toxic to the cell, MMP inhibits neither IPP nor squalene synthesis. This was not unexpected since previous work demonstrated the exclusion of polar compounds from intact cells (14). In contrast to the lack of passive transport mechanisms for highly polar salts, hepatocytes possess an active transport system for the taurine conjugates of bile acids. Because of the structural similarities between sulfonic and phosphonic acids, we thought that active transport of appropriately conjugated inhibitors might occur *via* the bile salt transport system. To test this idea, MMP was coupled to a reduced form of cholic acid through the carboxylic acid function of MMP. This preserved the structure of the inhibitor and produced a structural analog of taurocholic acid in which the sulfonate function was replaced by a phosphonic acid.

The effect of conjugation on deliver of MMP to the liver was examined by measuring the amount of [^{14}C]-MMP found in the various tissues and fluid spaces of the enterohepatic circulation. Rats received [^{14}C]-MMP, either free or conjugated to cholyl alcohol, by injection through a jugular vein catheter. The fraction of the dose found 60 min later in the blood, liver, the small and large intestine, and the intestinal contents are shown in Table 3. Conjugation

Table 3: Distributions of MMP and Cholyl-MMP in Selected Tissues and Fluids of the Rat.

Organ	Percent of Dose		
	MMP	Cholyl-MMP	Taurocholic Acid
Experiment 1			
Blood	1.4	8.3	-
Liver	5.3	26.0	-
Small Intestine	3.1	10.6	-
Large Intestine	3.1	6.4	-
Intestinal contents	3.5	8.6	-
Experiment 2			
Blood	2.0	<1.0	<1.0
Bile	1.2	10.0	38

Rats were fitted with jugular vein catheters (and bile-duct cannulae, Expt. 2) under light ether anesthesia. ^{14}C-Labeled test compounds, dissolved in saline, were administered through the catheters and the rats were maintained under light anesthesia for another 60 min. Then they were exsanguinated, and tissues were collected and analyzed for ^{14}C content as described in Methods. Results are the mean determinations from two animals. Test compounds infused in experiment 1 were: [5-^{14}C]MMP, 40 nCi (4 nmol) and cholyl-[15-^{14}C]MMP, 23.3 nCi (2.33 nmol); and in experiment 2: [5-^{14}C]MMP, 21.6 nCi (2.16 nmol); cholyl-[5-^{14}C]MMP, 23.3 nCi (2.33 nmol); and [24-^{14}C]taurocholic acid, 217 nCi (4.34 nmol).

to cholyl alcohol increased the hepatic uptake of MMP from 5.3% to 26% of the dose and increased intestinal uptake from 9.7% to 25.6% so that delivery to the organs of the enterohepatic circulation was increased from 16.5% to 60%. Most of the remaining dose was found in the bladder, urine, and carcass.

In order to determine how much of the increased delivery to the intestine was due to hepatic and biliary secretion, the experiment was repeated in animals having common bile duct cannulae. Bile was collected for 60 min following administration of taurocholic acid or [^{14}C]-MMP, either free or conjugated to cholyl alcohol, through a jugular vein catheter. All compounds were cleared rapidly from the blood (90% within 20 min). However, only taurocholic acid and MMP conjugated to cholyl alcohol, but not free MMP, were secreted into bile (Table 3).

Inhibition of Cholesterol Synthesis in Rat Hepatocytes

The effect of free and conjugated MMP on the rate of cholesterol synthesis from [2-^{14}C]-mevalonate by rat hepatocytes are shown in Table 4. Unconjugated MMP and taurocholic acid did not inhibit cholesterol synthesis either immediately or after a 3 h preincubation. In contrast, both the phosphonate diethyl ester and the free acid forms of MMP strongly inhibited cholesterol synthesis when they were presented to hepatocytes as the cholyl conjugates.

Table 4: Inhibition of Cholesterol Synthesis from *RS*-[5-^{3}H]mevalonic Acid by Rat Hepatocytes

Additions	Rate of Cholesterol Synthesis (nCi/h mg cell protein)	
	No Preincubation	Preincubation
None	50±6	51±4
Taurocholic acid	42±8	47±15
MMP	52±9	53±7
Cholyl-MMP	27±5	27±7
Cholyl-diethyl MMP	26±3	12±7

Hepatocytes were prepared from chow-fed rats and plated in DME containing 20% FCS for 24 h as described in Methods. Three h after changing the medium to serum-free DME, the medium was replaced with serum-free DME containing test compounds at a final concentration of 1 mM. Incubations were then started by addition of *RS*-[5-^{3}H]mevalonic acid (final concentration, 1 uM), either immediately or following a 3 h preincubation. Results are presented as the + SD (n=3).

Preincubation increased the degree of inhibition by the conjugated diethyl ester form.

DISCUSSION

MMP is a potent inhibitor of phosphomevalonate phosphokinase (4) and its ability to inhibit the synthesis of both IPP and squalene in the rat S_{10} system was therefore expected. The exclusion of charged molecules from transport into cells is a general problem. However, in recent years many investigations have described efficient, carrier-mediated transport of a variety of highly polar bile salts and their glycine or taurine conjugates into liver in whole animals, perfused liver, or hepatocytes (15-24). Moreover, others have demonstrated that analogs of the bile acid 3-sulfates (25) are taken up by liver. Thus we postulated that phosphonic acid derivatives resembling the natural amino acid sulfonate taurine, might be taken up by hepatocytes if coupled to the side chain of a carrier bearing the fundamental cholic acid ring structure. This was accomplished by reduction of cholic acid to the tetrol compound (IV) and its coupling specifically through the primary hydroxyl of the side chain to the free carboxyl of the phosphonate diester (II). Once the phosphonate diester linkages were removed, the remaining conjugate bore a significant structural similarity to taurocholic acid, including the complete ring chain system, three secondary hydroxyls, and an acidic function on the side chain.

Coupling MMP to reduced cholic acid provided an effective means of introducing an inhibitor of cholesterol synthesis into liver cells, albeit in a masked form. Conjugated MMP was not a good inhibitor of IPP synthesis in the S_{10} system (data not shown), but did inhibit cholesterol synthesis in hepatocytes. This together with the enhancement of inhibition brought about by preincubation of hepatocytes with conjugated diethyl MMP suggests that free MMP might be generated in hepatocytes. Our data reported here leave open many questions regarding the metabolism of bile salt analogs after they are taken up into hepatocytes. Further studies are needed to characterize the metabolic fate of MMP and other conjugates and to determine the general usefulness of bile salt analogs as carriers for metabolic probes and biologically active agents.

ACKNOWLEDGMENTS

This work was supported by grants from the National Institutes of Health (HL-24190 to TSP) and the PSC-BHE Research Award Program (RF 13354 to RE and RF 63330 to BET).

REFERENCES

1. Engel, R. (1983). **In**: *The Role of Phosphonates in Living Systems*, (Hilderbrand, R. L. Ed.) CRC Press, Inc., Boca Raton, FL, Chapter 5.
2. Popjak, G. (1970). *Ann. Intern Med.*, **72**:106-109.
3. Corey, E. J., and Volante, R. P. (1976). *J. Am. Chem. Soc.*, **98**:1291-1293.
4. Popjak, G., Parker, T. S., Sarin, V., Tropp, B. E., and Engel, R. (1978). *J. Am. Chem. Soc.*, **100**:8014-8016.
5. Parker, T. S., Popjak, G., Sutherland, K., and Wong, S.-M. (1978). *Biochem. Biophys. Acta*, **530**:24-34.
6. Sarin, V. K., Tropp, B. E., and Engel, R. (1977). *Tetrahedron Lett.*, 351-354.
7. Engel, R., Sarin, V. K., Gotlinsky, B., Tropp, B. E., and Parker, T. S., U.S. 4,279,989, 7/21/81.
8. Popjak, G. (1969). **In**: *Methods in Enzymology, XV*, (Clayton, B., Ed.) Academic Press, New York, NY, 394-454.
9. Gosselin, L., and Popjak, G. (1959). *Biochem. J.*, **71**:660-665.
10. Bucher, N., and McGarrahan, K. (1956). *J. Biol. Chem.*, **222**:1-15.
11. Berry, N. M., and Friend D. S. (1969). *J. Cell Biol.*, **43**:506-520.
12. Edwards, P. A. (1975). *Arch. Biochem. Biophys.*, **170**:188-203.
13. Cristophe, J., and Popjak, G. (1961). *J. Lipid Res.*, **2**:244-257.
14. Goldstein, S. L., Pulcrano, M., Tropp, B. E., and Engel, R. (1974). *J. Med. Chem.*, **17**:1115-1117.
15. O'Maille, E. R. L., Richards, T. G., and Short, A. H. (1967). *J. Physiol. Lond.*, **189**:337-350.
16. Glasinovic, J. C., Dumont M., Duval, M., and Erlinger S. (1975). *J. Clin. Invest.*, **55**:419-426.
17. Schwarz, L. R., Burr, R., Schwenk, M. Pfall, E., and Greim, H. (1975). *Eur. J. Biochem.*, **55**:617-623.
18. Anwer, M. S. R.,Kroker, R., and Hegner, D. (1976). *Biochem. Biophys. Res. Commun.*, **73**:63-71.
19. Reichen, J., and Paumgartner, G. (1976). *Am. J. Physiol.*, **231**:734-742.
20. Anwer, M. S., Hegner, D. (1978). *Hoppe-Seyler's, Z. Physiol.* Chem., **359**:181-192.

21. Pries, J. M., Sherman, C. A., Williams, G. C., and Hansoon, R. F. (1979). *Am. J. Physiol.*, **236**:E191-E197.
22. Iga, T., and Klaassen, C. D. (1982). *Biochem. Pharmacol.*, **31**:205-209, and 211-216.
23. Hoffman, N. E., Isler, J. H., and Smallwood, R. A. (1975). *Am J. Physiol.*, **229**298-302.
24. Ho, N. F., Mahjour, M., and Park, J. Y. (1982). **In:** *Alfred Benson Symposium 17*, (Buudgaard, H., Hanson, A. B., and Kofod, H., Eds.) Munksgaard, Copenhagen, pp. 80-92.
25. Spenney, J. C., Tobin, M. M., Mihas, A. A., Gibson, R. G.,Hirschowitz, B. I., Johnson, B. J., and Tauxe, W. N. (1979). *Gastroenterology*, **76**:272-278.

Index

A

B

C

D

E

F

G

H

I

K

L

M